Classic Texts Series

THE CONTEST PROBLEM

BOOK I

Annual High School Mathematics Examinations 1950-1960

Charles T. Salkind

arihant

ARIHANT PRAKASHAN (Series), MEERUT

arihant

ARIHANT PRAKASHAN (Series), MEERUT

卐 **Administrative & Production Offices**

Regd. Office

'Ramchhaya' 4577/15, Agarwal Road, Darya Ganj, New Delhi -110002

Tele: 011- 47630600, 43518550

卐 **Head Office**

Kalindi, TP Nagar, Meerut (UP) - 250002

Tel: 0121-7156203, 7156204

卐 **Sales & Support Offices**

Agra, Ahmedabad, Bengaluru, Bareilly, Chennai, Delhi, Guwahati, Hyderabad, Jaipur, Jhansi, Kolkata, Lucknow, Nagpur & Pune.

卐 ISBN 978-93-24191-76-2

Published by Arihant Publications (India) Ltd.

For further information about the books published by Arihant, log on to www.arihantbooks.com or e-mail at info@arihantbooks.com

Follow us on

Preface

Mathematical problems are older than mathematics itself. Stated as puzzles, they are found in the oldest written records. An Egyptian papyrus, dating back to 2200 B.C., asks: "A heap and its seventh make 19, how large is the heap?" Although' this is stated almost as tersely as we would do it today ("if $x + (1/7)x = 19$, find x"), ancient problems often appear in more poetical garb. Here is a charming example from India (recorded about 1150 A.D.): Out of a heap of pure lotus flowers, a third part, a fifth and a. sixth were offered respectively to the gods Siva, Vishnu, and then Sun; a quarter of the original heap was presented to Bhavani, The remaining six lotuses were given to the venerable preceptor. Tell quickly the whole number of lotus flowers. This sounds more exciting than: Solve the equation

$$\frac{1}{3}x + \frac{1}{5}x + \frac{1}{6}x + \frac{1}{4}x + 6 = x;$$

but of course our way of stating the problem has an overriding advantage : it makes the solution easier.

The glory of elevating mathematics from the level of problem solving to that of a science belongs to the ancient Greeks. But these Greek mathematicians did not neglect the art of inventing and solving problems. Three of their proposed geometric problems remained challenges to mathematicians until the 19th century. The "three famous problems of antiquity" asked for the trisection of an angle, the doubling of a cube, and the squaring of a circle, using only straight edge and compass. In spite of centuries of failure with these problems, efforts continued until it was proved, with the aid of algebraic theorems, that, in each case, the construction is impossible with the specified tools.

Since the time of the Greeks, such problems have stimulated the growth of mathematics and led mathematicians to invent new methods and to develop new concepts. On a more modest but equally important level, such problems have provided an unexcelled training ground for young mathematicians. But problem-solving never was the exclusive domain of professionals. At all times it provided intellectual stimulation and joy for many professionally outside the field of mathematics.

During the Renaissance in Italy problem-solving became a competitive sport and public contests between mathematicians were not uncommon. Sometimes the competitive spirit led to excesses. For example, in the 16th century, Tartaglia, the winner of such a contest, had to flee town to escape violence at the hands of the fans of the local champion. The subject of this particular contest was cubic equations and, in preparing for it, Tartaglia discovered important formulas for solving them. In our own day mathematicians still compete with each other in solving problems-less publicly, less violently, but perhaps with the same intensity.

The end of the 19th century saw the beginning of organized competitions among secondary school students. The so-called Eotvos Competition in Hungary (begun 'in 1894) is justly famous; it has probably played its part in producing so many superior mathematicians and physicists from this small country. In the Soviet Union high school students take part in a university-sponsored system of mathematical "olympics".

In America the tradition of periodic competitions among students of mathematics is almost fifty years old. There are, for example, the Inter-scholastic Mathematics League in New York City and Long Island and the annual competitions conducted by Pi Mu Epsilon of New York University; there are also state-wide and regional programs in Texas, Wisconsin, Indiana, and at Stanford University, and elsewhere. As our first problem book we present the complete set of problems appearing in the annual examinations.

The MAA is concerned primarily with mathematics on the undergraduate level. It is one of the three major mathematical organizations in America (the other two being the American Mathematical . Society, chiefly concerned with mathematical research, and the National Council of Teachers of Mathematics, concerned with the content and pedagogy of elementary and secondary mathematics). The MAA also conducts the annual "Putnam Competition" for under-graduate students. Its journal, The American Mathematical Monthly, is famous for its elementary and advanced sections.

When the annual examination was first organized in 1950 it was restricted to the Metropolitan New York area. It became a national project in 1957, receiving in the same year co-sponsorship by the Society of Actuaries. In 1960 more than 150,000 students from 5,200 schools participated in the program. The examination is conducted in nearly every state and territory of the U.S. and in the more-populated provinces of Canada.

The annual examination is based entirely on the standard high school curriculum (except for a very few problems in the earlier tests) and pre-suppose a no advanced knowledge. Readers who find these problems too easy are advised to try the problem sections of other NML books and to await additional NML problem books which are now in preparation.

Part I of each examination tests fundamental skills based on conceptual understanding, while. Parts II and In probe beyond mere reproduction of class-room work. For the ten years this examination was conducted, only three people scored perfectly (150); a score of 80 or more places a contestant on the Honor Roll

The editors are grateful to the MAA for permission to publish this collection, and to Prof. Charles T. Salkind for compiling the book and for supplying a

classification of problems with their complete solutions. In preparing this collection, the compiler and the editors have made a few minor changes in the statements of the original examination problems, for the sake of greater clarity.

This problem collection is designed to be used by mathematics clubs, high school teachers, students, and other interested individuals. Clearly, no one would profit from doing all the problems, but he would benefit from those that present a challenge to him. The reader might try himself on a whole test or on part of a test, with (or preferably without) time limitations.

He should try to get as far as possible with the solution to a problem. If he is really stuck, he should look up the answer in the key, see pages 75-78, and try to work backwards; if this fails, the section of complete solutions should be consulted, see pages 79-149.

In studying solutions, even the successful problem solver -may find sidelights he had overlooked; he may find a more "elegant" or a different way of solving the problem which may lead him deeper into mathematics. He may find it interesting to change items in the hypothesis and to see how this affects .the solution, or to invent his own problems.

If a reader is interested in a special type of problem, he should consult the classified index.

List of Symbols

Symbol	Meaning
$\sim$	similar (if used in connection with plane figures)
$\sim$	approximately equal (if used in connection with numbers)
$\therefore$	therefore
$\equiv$	identically equal to
$<$	less than
$\leq$	less than or equal to
$>$	greater than
$\geq$	greater than or equal to
$\lvert k \rvert$	absolute value of the number k
Δ	triangle
110_2	the number $12^2 + 1.2^1 + 0.2^0$, i.e., the number 6 when written in a number system with base 2 instead of 10.
$\cong$	congruent
$\neq$	different from
$\perp$	perpendicular to
$\overline{XY}$	length of the line segment XY

Contents

I Problems

1950 Examination 3
1951 Examination 11
1952 Examination 20
1953 Examination 29
1954 Examination 37
1955 Examination 46
1956 Examination 55
1957 Examination 63
1958 Examination 72
1959 Examination 81
1960 Examination 89

II Answer Keys **99**

III Solutions

1950 Solutions 103
1951 Solutions 111
1952 Solutions 117
1953 Solutions 125
1954 Solutions 131
1955 Solutions 138
1956 Solutions 145
1957 Solutions 152
1958 Solutions 158
1959 Solutions 165
1960 Solutions 172

IV Classification of Problems **179**

PROBLEMS

I
Problems

1950 Examination

Part 1

1. If 64 is divided into three parts proportional to 2, 4 and 6, the smallest part is :

(A) $5\frac{1}{3}$ (B) 11 (C) $10\frac{2}{3}$

(D) 5 (E) none of these answers

2. Let $R = gS - 4$. When $S = 8$, $R = 16$. When $S = 10$, R is equal to:

(A) 11 (B) 14 (C) 20

(D) 21 (E) none of these

3. The sum of the roots of the equation $4x^2 + 5 - 8x = 0$ is equal to:

(A) 8 (B) -5 (C) $-\frac{5}{4}$

(D) -2 (E) none of these

4. Reduced to lowest terms, $\frac{a^2 - b^2}{ab} - \frac{ab - b^2}{ab - a^2}$ is equal to:

(A) $\frac{a}{b}$ (B) $\frac{a^2 - 2b^2}{ab}$ (C) a^2

(D) $a - 2b$ (E) none of these

5. If five geometric means are inserted between 8 and 5832, the fifth term in the geometric series is:

(A) 648 (B) 832 (C) 1168

(D) 1944 (E) none of these

6. The values of y which will satisfy the equations

$2x^2 + 6x + 5y + 1 = 0,\ 2x + y + 3 = 0$ may be found by solving:

(A) $y^2 + 14y - 7 = 0$ (B) $y^2 + 8y + 1 = 0$

(C) $y^2 + 10y - 7 = 0$ (D) $y^2 + y - 12 = 0$

(E) none of these equations

7. If the digit 1 is placed after a two digit number whose tens' digit is t, and units' digit is u, the new number is:

(A) $10t + u + 1$ (B) $100t + 10u + 1$ (C) $1000t + 10u + 1$
(D) $t + u + 1$ (E) none of these answers

8. If the radius of a circle is increased 100%, the area is increased:

(A) 100% (B) 200% (C) 300%
(D) 400% (E) by none of these

9. The area of the largest triangle that can be inscribed in a semi-circle whose radius is r is:

(A) r^2 (B) r^3 (C) $2r^2$
(D) $2r^3$ (E) $\frac{1}{2}r^2$

10. After rationalizing the numerator of $\frac{\sqrt{3}-\sqrt{2}}{\sqrt{3}}$, the denominator in simplest form is:

(A) $\sqrt{3}(\sqrt{3}+\sqrt{2})$ (B) $\sqrt{3}(\sqrt{3}-\sqrt{2})$ (C) $3-\sqrt{3}\sqrt{2}$
(D) $3+\sqrt{6}$ (E) none of these answers

11. In the formula $C = \frac{en}{R+nr}$, e, R and r are positive constants, and n is a positive integer. As n increases, C:

(A) decreases (B) increases (C) remains constant
(D) increases and then decreases (E) decreases and then increases

12. As the number of sides of a polygon increases from 3 to n, the sum of the exterior angles formed by extending each side in succession:

(A) increases (B) decreases (C) remains constant
(D) cannot be predicted
(E) becomes $(n-3)$ straight angles

13. The roots of $(x^2-3x+2)(x)(x-4)=0$ are:

(A) 4 (B) 0 and 4 (C) 1 and 2
(D) 0, 1, 2 and 4 (E) 1, 2 and 4

14. For the simultaneous equations $2x-3y=8$, $6y-4x=9$

(A) $x=4,\ y=0$ (B) $x=0,\ y=\frac{3}{2}$ (C) $x=0,\ y=0$
(D) there is no solution
(E) there are an infinite number of solutions

15. The real factors of x^2+4 are:

(A) $(x^2+2)(x^2+2)$ (B) $(x^2+2)(x^2-2)$ (C) $x^2(x^2+4)$
(D) $(x^2-2x+2)(x^2+2x+2)$ (E) non-existent

Part 2

16. The number of terms in the expansion of $[(a+3b)^2(a-3b)^2]^2$ when simplified is:

(A) 4 (B) 5 (C) 6

(D) 7 (E) 8

17. The formula which expresses the relationship between x and y as shown in the accompanying table is:

x	0	1	2	3	4
y	100	90	70	40	0

(A) $y=100-10x$ (B) $y=100-5x^2$ (C) $y=100-5x-5x^2$

(D) $y=20-x-x^2$ (E) none of these

18. Of the following

(1) $a(x-y)=ax-ay$ (2) $a^{x-y}=a^x-a^y$

(3) $\log(x-y)=\log x-\log y$ (4) $\log x/\log y=\log x-\log y$

(5) $a(xy)=ax\cdot ay$:

(A) only 1 and 4 are true (B) only 1 and 5 are true

(C) only 1 and 3 are true (D) only 1 and 2 are true

(E) only 1 is true

19. If m men can do a job in d days, then $m+r$ men can do the job in:

(A) $d+r$ days (B) $d-r$ days (C) $\frac{md}{m+r}$ days

(D) $\frac{d}{m+r}$ days (E) none of these

20. When $x^{13}+1$ is divided by $x-1$, the remainder is:

(A) 1 (B) -1 (C) 0

(D) 2 (E) none of these answers

21. The volume of a rectangular solid each of whose side, front, and bottom faces are 12 sq. in., 8 sq. in., and 6 sq. in. respectively is:

(A) 576 cu. in. (B) 24 cu. in. (C) 9 cu. in.

(D) 104 cu. in. (E) none of these

22. Successive discounts of 10% and 20% are equivalent to a single discount of:

(A) 30% (B) 15% (C) 72%

(D) 28% (E) none of these

23. A man buys a house for \$10,000 and rents it. He puts $12\frac{1}{2}\%$ of each month's rent aside for repairs and upkeep; pays \$325 a year taxes and realizes $5\frac{1}{2}\%$ on his investment. The monthly rent is:

(A) \$64.82 (B) \$83.33 (C)\$72.08
(D) \$45.83 (E) \$177.08

24. The equation $x + \sqrt{x-2} = 4$ has:

(A) 2 real roots (B) 1 real and 1 imaginary root
(C) 2 imaginary roots (D) no roots
(E) 1 real root

25. The value of $\log_5 \frac{(125)(625)}{25}$ is equal to:

(A) 725 (B) 6 (C) 3125
(D) 5 (E) none of these

26. If $\log_{10} m = b - \log_{10} n$, then $m =$

(A) $\frac{b}{n}$ (B) bn (C) $10^b n$
(D) $b - 10^n$ (E) $\frac{10^b}{n}$

27. A car travels 120 miles from A to B at 30 miles per hour but returns the same distance at 40 miles per hour. The average speed for the round trip is closest to:

(A) 33 mph (B) 34 mph (C) 35 mph
(D) 36 mph (E) 37 mph

28. Two boys A and B start at the same time to ride from Port Jervis to Poughkeepsie, 60 miles away. A travels 4 miles an hour slower than B. B reaches Poughkeepsie and at once turns back meeting A 12 miles from Poughkeepsie. The rate of A was:

(A) 4 mph (B) 8 mph (C) 12 mph
(D) 16 mph (E) 20 mph

29. A manufacturer built a machine which will address 500 envelopes in 8 minutes. He wishes to build another machine so that when both are operating together they will address 500 envelopes in 2 minutes. The equation used to find how many minutes x it would require the second machine to address 500 envelopes alone is:

(A) $8 - x = 2$ (B) $\frac{1}{8} + \frac{1}{x} = \frac{1}{2}$ (C) $\frac{500}{8} + \frac{500}{x} = 500$
(D) $\frac{x}{2} + \frac{x}{8} = 1$ (E) none of these answers

30. From a group of boys and girls, 15 girls leave. There are then left two boys for each girl. After this 45 boys leave. There are then 5 girls for each boy. The number of girls in the beginning was:

(A) 40 (B) 43 (C) 29
(D) 50 (E) none of these

31. John ordered 4 pairs of black socks and some additional pairs of blue socks. The price of the black socks per pair was twice that of the blue. When the order was filled, it was found that the number of pairs of the two colors had been interchanged. This increased the bill by 50%. The ratio of the number of pairs of black socks to the number of pairs of blue socks in the original order was:

(A) 4 : 1 (B) 2 : 1 (C) 1 : 4
(D) 1 : 2 (E) 1 : 8

32. A 25 foot ladder is placed against a vertical wall of a building. The foot of the ladder is 7 feet from the base of the building. If the top of the ladder slips 4 feet, then the foot of the ladder will slide:

(A) 9 ft. (B) 15 ft. (C) 5 ft.
(D) 8 ft. (E) 4 ft.

33. The number of circular pipes with an inside diameter of 1 inch which will carry the same amount of water as a pipe with an inside diameter of 6 inches is:

(A) 6π (B) 6 (C) 12
(D) 36 (E) 36π

34. When the circumference of a toy balloon is increased from 20 inches to 25 inches, the radius is increased by:

(A) 5 in. (B) $2\frac{1}{2}$ in. (C) $5/\pi$ in.
(D) $5/2\pi$ in. (E) $\pi/5$ in.

35. In ΔABC, $\overline{AC}=24''$, $\overline{BC}=10''$, and $\overline{AB}=26''$. The radius of the inscribed circle is:

(A) 26 in. (B) 4 in. (C) 13 in.
(D) 8 in. (E) none of these

Part 3

36. A merchant buys goods at 25% off the list price. He desires to mark the goods so that he can give a discount of 20% on the marked price and still clear a profit of 25% on the selling price. What per cent of the list price must he mark the goods?

(A) 125% (B) 100% (C) 120%
(D) 80% (E) 75%

37. If $y = \log_a x$, and $a > 1$, which of the following statements is incorrect?

(A) if $x = 1$, $y = 0$ (B) if $x = a$, $y = 1$

(C) if $x = -1$, y is imaginary (complex)

(D) if $0 < x < 1$, y is always less than 0 and decreases without limit as x approaches zero

(E) only some of the above statements are correct

38. If the expression $\begin{vmatrix} a & c \\ d & b \end{vmatrix}$ has the value $ab - cd$ for all values of a, b, c and d, then the equation $\begin{vmatrix} 2x & 1 \\ x & x \end{vmatrix} = 3$:

(A) is satisfied for only 1 value of x

(B) is satisfied for 2 values of x

(C) is satisfied for no values of x

(D) is satisfied for an infinite number of values of x

(E) none of these

39. Given the series $2 + 1 + \frac{1}{2} + \frac{1}{4} + \ldots$ and the following five statements:

(1) the sum increases without limit.

(2) the sum decreases without limit.

(3) the difference between any term of the sequence and zero can be made less than any positive quantity no matter how small.

(4) the difference between the sum and 4 can be made less than any positive quantity no matter how small.

(5) the sum approaches a limit.

Of these statements, the correct ones are:

(A) only (3) and (4) (B) only (5) (C) only (2) and (4)

(D) only (2), (3) and (4) (E) only (4) and (5)

40. The limit of $\frac{x^2 - 1}{x - 1}$ as x approaches 1 as a limit is:

(A) 0 (B) indeterminate (C) $x - 1$

(D) 2 (E) 1

41. The least value of the function $ax^2 + bx + c (a > 0)$ is:

(A) $-\frac{b}{a}$ (B) $-\frac{b}{2a}$

(C) $b^2 - 4ac$ (D) $\frac{4ac - b^2}{4a}$

(E) none of these

42. The equation $x^{x^{x^{\cdot^{\cdot^{\cdot}}}}} = 2$ is satisfied when x is equal to:

(A) infinity (B) 2 (C) $\sqrt[4]{2}$

(D) $\sqrt{2}$ (E) none of these

43. The sum to infinity of $\frac{1}{7} + \frac{2}{7^2} + \frac{1}{7^3} + \frac{2}{7^4} + \dots$ is:

(A) $\frac{1}{5}$ (B) $\frac{1}{24}$

(C) $\frac{5}{48}$ (D) $\frac{1}{16}$

(E) none of these

44. The graph of $y = \log x$

(A) cuts the y-axis

(B) cuts all lines perpendicular to the x-axis

(C) cuts the x-axis

(D) cuts neither axis

(E) cuts all circles whose centre is at the origin

45. The number of diagonals that can be drawn in a polygon of 100 sides is:

(A) 4850 (B) 4950 (C) 9900

(D) 98 (E) 8800

46. In triangle ABC, $\overline{AB} = 12$, $\overline{AC} = 7$, and $\overline{BC} = 10$. If sides AB and AC are doubled while BC remains the same, then:

(A) the area is doubled

(B) the altitude is doubled

(C) the area is four times the original area

(D) the median is unchanged

(E) the area of the triangle is 0

47. A rectangle inscribed in a triangle has its base coinciding with the base b of the triangle. If the altitude of the triangle is h, and the altitude x of the rectangle is half the base of the rectangle, then:

(A) $x = \frac{1}{2}h$ (B) $x = \frac{bh}{h+b}$

(C) $x = \frac{bh}{2h+b}$ (D) $x = \sqrt{\frac{hb}{2}}$

(E) $x = \frac{1}{2}b$

48. A point is selected at random inside an equilateral triangle. From this point perpendiculars are dropped to each side. The sum of these perpendicular is:

(A) least when the point is the centre of gravity of the triangle

(B) greater than the altitude of the triangle

(C) equal to the altitude of the triangle

(D) one-half the sum of the sides of the triangle

(E) greatest when the point is the center of gravity

49. A triangle has a fixed base AB that is 2 inches long. The median from A to side BC is $1\frac{1}{2}$ inches long and can have any position emanating from A. The locus of the vertex C of the triangle is:

(A) a straight line AB, $1\frac{1}{2}$ in. from A

(B) a circle with A as center and radius 2 in.

(C) a circle with A as center and radius 3 in.

(D) a circle with radius 3 in. and center 4 in. from B along BA

(E) an ellipse with A as a focus

50. A privateer discovers a merchantman 10 miles to leeward at 11:45 a.m. and with a good breeze bears down upon her at 11 mph, while the merchantman can only make 8 mph in her attempt to escape. After a two hour chase, the top sail of the privateer is carried away, she can now make only 17 miles while the merchantman makes 15. The privateer will overtake the merchantman at:

(A) 3:45 p.m. (B) 3:30 p.m. (C) 5:00 p.m.

(D) 2:45 p.m. (E) 5:30 p.m.

1951 Examination

Part 1

1. The per cent that M is greater than N, is:

(A) $\frac{100(M-N)}{M}$ (B) $\frac{100(M-N)}{N}$ (C) $\frac{M-N}{N}$

(D) $\frac{M-N}{M}$ (E) $\frac{100\,(M+N)}{N}$

2. A rectangular field is half as wide as it is long and is completely enclosed by x yards of fencing. The area in terms of x is:

(A) $\frac{x^2}{2}$ (B) $2x^2$ (C) $\frac{2x^2}{9}$

(D) $\frac{x^2}{18}$ (E) $\frac{x^2}{72}$

3. If the length of a diagonal of a square is $a + b$, then the area of the square is:

(A) $(a+b)^2$ (B) $\frac{1}{2}(a+b)^2$ (C) a^2+b^2

(D) $\frac{1}{2}(a^2+b^2)$ (E) none of these

4. A barn with a flat roof is rectangular in shape, 10 yd. wide, 13 yd. long and 5 yd. high. It is to be painted inside and outside, and on the ceiling, but not on the roof or floor. The total number of sq. yd. to be painted is:

(A) 360 (B) 460 (C) 490

(D) 590 (E) 720

5. Mr. A owns a home worth \$10,000. He sells it to Mr. B at a 10% profit based on the worth of the house. Mr. B sells the house back to Mr. A at a 10% loss. Then:

(A) A comes out even (B) A makes \$1100 on the deal

(C) A makes \$1000 on the deal (D) A loses \$900 on the deal

(E) A loses \$1000 on the deal

6. The bottom, side, and front areas of a rectangular box are known. The product of these areas is equal to:

(A) the volume of the box (B) the square root of the volume
(C) twice the volume (D) the square of the volume
(E) the cube of the volume

7. An error of .02″ is made in the measurement of a line 10″ long, while an error of only .2″ is made in a measurement of a line 100″ long. In comparison with the relative error of the first measurement, the relative error of the second measurement is:

(A) greater by .18 (B) the same (C) less
(D) 10 times as great (E) correctly described by both (A) and (D)

8. The price of an article is cut 10%. To restore it to its former value, the new price must be increased by:

(A) 10% (B) 9% (C) $11\frac{1}{9}\%$
(D) 11% (E) none of these answers

9. An equilateral triangle is drawn with a side of length a. A new equilateral triangle is formed by joining the mid-points of the sides of the first one. Then a third equilateral triangle is formed by joining the mid-points of the sides of the second, and so on forever. The limit of the sum of the perimeters of all the triangles thus drawn is:

(A) Infinite (B) $5\frac{1}{4}a$ (C) $2a$
(D) $6a$ (E) $4\frac{1}{2}a$

10. Of the following statements, the one that is incorrect is:

(A) Doubling the base of a given rectangle doubles the area.
(B) Doubling the altitude of a triangle doubles the area.
(C) Doubling the radius of a given circle doubles the area.
(D) Doubling the divisor of a fraction and dividing its numerator by 2 changes the quotient.
(E) Doubling a given quantity may make it less than it originally was.

11. The limit of the sum of an infinite number of terms in a geometric progression is $a/(1-r)$ where a denotes the first term and $-1 < r < 1$ denotes the common ratio. The limit of the sum of their squares is:

(A) $\dfrac{a^2}{(1-r)^2}$ (B) $\dfrac{a^2}{1+r^2}$ (C) $\dfrac{a^2}{1-r^2}$
(D) $\dfrac{4a^2}{1+r^2}$ (E) none of these

12. At 2:15 o'clock, the hour and minute hands of a clock form an angle of:

(A) $30°$ (B) $5°$ (C) $22\frac{1}{2}°$

(D) $7\frac{1}{2}°$ (E) $28°$

13. *A* can do a piece of work in 9 days. *B* is 50% more efficient than *A*. The number of days it takes *B* to do the same piece of work is:

(A) $13\frac{1}{2}$ (B) $4\frac{1}{2}$ (C) 6

(D) 3 (E) none of these answers

14. In connection with proof in geometry, indicate which one of the following statements is *incorrect*:

(A) Some statements are accepted without being proved.

(B) In some instances there is more than one correct order in proving certain propositions.

(C) Every term used in a proof must have been defined previously.

(D) It is not possible to arrive by correct reasoning at a true conclusion if, in the given, there is an untrue proposition.

(E) Indirect proof can be used whenever there are two or more contrary propositions.

15. The largest number by which the expression $n^3 - n$ is divisible for all possible integral values of n, is:

(A) 2 (B) 3 (C) 4

(D) 5 (E) 6

Part 2

16. If in applying the quadratic formula to a quadratic equation $f(x) \equiv ax^2 + bx + c = 0$, it happens that $c = b^2/4a$, then the graph of $y = f(x)$ will certainly:

(A) have a maximum (B) have a minimum

(C) be tangent to the x-axis (D) be tangent to the y-axis

(E) lie in one quadrant only

17. Indicate in which one of the following equations y is neither directly nor inversely proportional to x:

(A) $x + y = 0$ (B) $3xy = 10$ (C) $x = 5y$

(D) $3x + y = 10$ (E) $\frac{x}{y} = \sqrt{3}$

18. The expression $21x^2 + ax + 21$ is to be factored into two linear prime binomial factors with integer coefficients. This can be done if a is:

(A) any odd number (B) some odd number
(C) any even number (D) some even number
(E) zero

19. A six place number is formed by repeating a three place number; for example, 256, 256, or 678, 678, etc. Any number of this form is always exactly divisible by:

(A) 7 only (B) 11 only (C) 13 only
(D) 101 (E) 1001

20. When simplified and expressed with negative exponents, the expression $(x+y)^{-1}(x^{-1}+y^{-1})$ is equal to:

(A) $x^{-2}+2x^{-1}y^{-1}+y^{-2}$ (B) $x^{-2}+2^{-1}x^{-1}y^{-1}+y^{-2}$
(C) $x^{-1}y^{-1}$ (D) $x^{-2}+y^{-2}$
(E) $\dfrac{1}{x^{-1}y^{-1}}$

21. Given: $x>0$, $y>0$, $x>y$ and $z \neq 0$. The inequality which is not always correct is:

(A) $x+z>y+z$ (B) $x-z>y-z$ (C) $xz>yz$
(D) $\dfrac{x}{z^2}>\dfrac{y}{z^2}$ (E) $xz^2>yz^2$

22. The values of a in the equation : $\log_{10}(a^2-15a)=2$ are :

(A) $\dfrac{15 \pm \sqrt{233}}{2}$ (B) $20, -5$ (C) $\dfrac{15 \pm \sqrt{305}}{2}$
(D) ± 20 (E) none of these

23. The radius of a cylindrical box is 8 inches and the height is 3 inches. The number of inches that may be added to either the radius or the height to give the same non-zero increase in volume is:

(A) 1 (B) $5\dfrac{1}{3}$ (C) any number
(D) non-existent (E) none of these

24. $\dfrac{2^{n+4}-2(2^n)}{2(2^{n+3})}$ when simplified is:

(A) $2^{n+1}-\dfrac{1}{8}$ (B) -2^{n+1} (C) $1-2^n$
(D) $\dfrac{7}{8}$ (E) $\dfrac{7}{4}$

25. The apothem of a square having its area numerically equal to its perimeter is compared with the apothem of an equilateral triangle having its area numerically equal to its perimeter. The first apothem will be:

(A) equal to the second
(B) $\frac{4}{3}$ times the second
(C) $\frac{2}{\sqrt{3}}$ times the second
(D) $\frac{\sqrt{2}}{\sqrt{3}}$ times the second
(E) indeterminately related to the second

26. In the equation $\frac{x(x-1)-(m+1)}{(x-1)(m-1)} = \frac{x}{m}$ the roots are equal when

(A) $m = 1$
(B) $m = \frac{1}{2}$
(C) $m = 0$
(D) $m = -1$
(E) $m = -\frac{1}{2}$

27. Through a point inside a triangle, three lines are drawn from the vertices to the opposite sides forming six triangular sections. Then:

(A) the triangles are similar in opposite pairs
(B) the triangles are congruent in opposite pairs
(C) the triangles are equal in area in opposite pairs
(D) three similar quadrilaterals are formed
(E) none of the above relations is true

28. The pressure (P) of wind on a sail varies jointly as the area (A) of the sail and the square of the velocity (V) of the wind. The pressure on a square foot is 1 pound when the velocity is 16 miles per hour. The velocity of the wind when the pressure on a square yard is 36 pounds is:

(A) $10\frac{2}{3}$ mph
(B) 96 mph
(C) 32 mph
(D) $1\frac{2}{3}$ mph
(E) 16 mph

29. Of the following sets of data the only one that does not determine the shape of a triangle is:

(A) the ratio of two sides and the included angle
(B) the ratios of the three altitudes
(C) the ratios of the three medians
(D) the ratio of the altitude to the corresponding base
(E) two angles

30. If two poles 20″ and 80″ high are 100″ apart, then the height of the intersection of the lines joining the top of each pole to the foot of the opposite pole is:

(A) 50″ (B) 40″ (C) 16″
(D) 60″ (E) none of these

31. A total of 28 handshakes was exchanged at the conclusion of a party. Assuming that each participant was equally polite toward all the others, the number of people present was:

(A) 14 (B) 28 (C) 56
(D) 8 (E) 7

32. If ΔABC is inscribed in a semicircle whose diameter is AB, then $\overline{AC}+\overline{BC}$ must be:

(A) equal to $\overline{AB}$ (B) equal to $\overline{AB}\sqrt{2}$ (C) $\geq AB\sqrt{2}$
(D) $\leq \overline{AB}\sqrt{2}$ (E) $\overline{AB}^2$

($a \geq b$ is read "a is equal to or greater than b.")

33. The roots of the equation $x^2-2x=0$ can be obtained graphically by finding the abscissas of the points of intersection of each of the following pairs of equations except the pair:

(A) $y=x^2,\ y=2x$ (B) $y=x^2-2x,\ y=0$
(C) $y=x,\ y=x-2$ (D) $y=x^2-2x+1,\ y=1$
(E) $y=x^2-1,\ y=2x-1$

34. The value of $10^{\log_{10}7}$ is:

(A) 7 (B) 1 (C) 10
(D) $\log_{10}7$ (E) $\log_7 10$

35. If $a^x=c^q=b$ and $c^y=a^z=d$, then:

(A) $xy=qz$ (B) $\frac{x}{y}=\frac{q}{z}$ (C) $x+y=q+z$
(D) $x-y=q-z$ (E) $x^y=q^z$

Part 3

36. Which of the following methods of proving a geometric figure a locus is not correct?

(A) Every point on the locus satisfies the conditions and every point not on the locus does not satisfy the conditions.

(B) Every point not satisfying the conditions is not on the locus and every point on the locus does satisfy the conditions.

(C) Every point satisfying the conditions is on the locus and every point on the locus satisfies the conditions.

(D) Every point not on the locus does not satisfy the conditions and every point not satisfying the conditions is not on the locus.

(E) Every point satisfying the conditions is on the locus and every point not satisfying the conditions is not on the locus.

37. A number which when divided by 10 leaves a remainder of 9, when divided by 9 leaves a remainder of 8, by 8 leaves a remainder of 7, etc., down to where, when divided by 2, it leaves a remainder of 1, is:

(A) 59 (B) 419 (C) 1259

(D) 2519 (E) none of these answers

38. A rise of 600 feet is required to get a railroad line over a mountain. The grade can be kept down by lengthening the track and curving it around the mountain peak. The additional length of track required to reduce the grade from 3% to 2% is approximately:

(A) 10,000 ft. (B) 20,000 ft. (C) 30,000 ft.

(D) 12,000 ft. (E) none of these

39. A stone is dropped into a well and the report of the stone striking the bottom is heard 7.7 seconds after it is dropped. Assume that the stone falls $16t^2$ feet in t seconds and that the velocity of sound is 1,120 feet per second. The depth of the well is:

(A) 784 ft. (B) 342 ft. (C) 1568 ft.

(D) 156.8 ft. (E) none of these

40. $\left(\dfrac{(x+1)^2(x^2-x+1)^2}{(x^3+1)^2}\right)^2 \cdot \left(\dfrac{(x-1)^2(x^2+x+1)^2}{(x^3-1)^2}\right)^2$ equals:

(A) $(x+1)^4$ (B) $(x^3+1)^4$ (C) 1

(D) $[(x^3+1)(x^3-1)]^2$ (E) $[(x^3-1)^2]^2$

41. The formula expressing the relationship between x and y in the table is:

x	2	3	4	5	6
y	0	2	6	12	20

(A) $y = 2x - 4$ (B) $y = x^2 - 3x + 2$

(C) $y = x^3 - 3x^2 + 2x$ (D) $y = x^2 - 4x$

(E) $y = x^2 - 4$

42. If $x = \sqrt{1+\sqrt{1+\sqrt{1+\sqrt{1+\ldots}}}}$, then:

(A) $x = 1$ (B) $0 < x < 1$

(C) $1 < x < 2$ (D) x is infinite

(E) $x > 2$ but finite

43. Of the following statements, the only one that is incorrect is:

(A) An inequality will remain true after each side is increased, decreased, multiplied or divided (zero excluded) by the same positive quantity.

(B) The arithmetic mean of two unequal positive quantities is greater than their geometric mean.

(C) If the sum of two positive quantities is given, their product is largest when they are equal.

(D) If a and b are positive and unequal, $\frac{1}{2}(a^2 + b^2)$ is greater than $\left[\frac{1}{2}(a + b)\right]^2$

(E) If the product of two positive quantities is given, their sum is greatest when they are equal.

44. If $\frac{xy}{x+y} = a$, $\frac{xz}{x+z} = b$ and $\frac{yz}{y+z} = c$, where a, b, and c are other than zero, then x equals:

(A) $\frac{abc}{ab+ac+bc}$ (B) $\frac{2abc}{ab+bc+ac}$ (C) $\frac{2abc}{ab+ac-bc}$

(D) $\frac{2abc}{ab+bc-ac}$ (E) $\frac{2abc}{ac+bc-ab}$

45. If you are given $\log 8 = .9031$ and $\log 9 = .9542$, then the only logarithm that cannot be found without the use of tables is:

(A) $\log 17$ (B) $\log(5/4)$ (C) $\log 15$

(D) $\log 600$ (E) $\log .4$

46. AB is a fixed diameter of a circle whose center is O. From C, any point on the circle,a chord CD is drawn perpendicular to AB. Then, as C moves over a semicircle, the bisector of angle OCD cuts the circle in a point that always:

(A) bisects the arc AB (B) trisects the arc AB

(C) varies (D) is as far from AB as from D

(E) is equidistant from B and C

47. If r and s are the roots of the equation $ax^2 + bx + c = 0$, the value of $\frac{1}{r^2} + \frac{1}{s^2}$ is:

(A) $b^2 - 4ac$ (B) $\frac{b^2 - 4ac}{2a}$ (C) $\frac{b^2 - 4ac}{c^2}$

(D) $\frac{b^2 - 2ac}{c^2}$ (E) none of these

48. The area of a square inscribed in a semicircle is to the area of the square inscribed in the entire circle as:

(A) $1:2$ (B) $2:3$ (C) $2:5$

(D) $3:4$ (E) $3:5$

49. The medians of a right triangle which are drawn from the vertices of the acute angles are 5 and $\sqrt{40}$. The value of the hypotenuse is:

(A) 10 (B) $2\sqrt{40}$ (C) $\sqrt{13}$

(D) $2\sqrt{13}$ (E) none of these

50. Tom, Dick and Harry started out on a 100-mile journey. Tom and Harry went by automobile at the rate of 25 mph, while Dick walked at the rate of 5 mph. After a certain distance, Harry got off and walked on at 5 mph, while Tom went back for Dick and got him to the destination at the same time that Harry arrived. The number of hours required for the trip was:

(A) 5 (B) 6 (C) 7

(D) 8 (E) none of these answers

1952 Examination

Part 1

1. If the radius of a circle is a rational number, its area is given by a number which is:

(A) rational (B) irrational (C) integral
(D) a perfect square (E) none of these

2. Two high school classes took the same test. One class of 20 students made an average grade of 80%; the other class of 30 students made an average grade of 70%. The average grade for all students in both classes is:

(A) 75% (B) 74% (C) 72%
(D) 77% (E) none of these

3. The expression $a^3 - a^{-3}$ equals:

(A) $\left(a-\frac{1}{a}\right)\left(a^2+1+\frac{1}{a^2}\right)$ (B) $\left(\frac{1}{a}-a\right)\left(a^2-1+\frac{1}{a^2}\right)$

(C) $\left(a-\frac{1}{a}\right)\left(a^2-2+\frac{1}{a^2}\right)$ (D) $\left(\frac{1}{a}-a\right)\left(\frac{1}{a^2}+1+a^2\right)$

(E) none of these

4. The cost C of sending a parcel post package weighing P pounds, P an integer, is 10 cents for the first pound and 3 cents for each additional pound. The formula for the cost is:

(A) $C = 10 + 3P$ (B) $C = 10P + 3$ (C) $C = 10 + 3(P-1)$
(D) $C = 9 + 3P$ (E) $C = 10P - 7$

5. The points (6, 12) and (0, – 6) are connected by a straight line. Another point on this line is:

(A) (3, 3) (B) (2, 1) (C) (7, 16)
(D) (– 1, – 4) (E) (– 3, – 8)

6. The difference of the roots of $x^2 - 7x - 9 = 0$ is:

(A) + 7 (B) $+\frac{7}{2}$ (C) + 9
(D) $2\sqrt{85}$ (E) $\sqrt{85}$

7. When simplified, $(x^{-1} + y^{-1})^{-1}$ is equal to:

(A) $x + y$ (B) $\frac{xy}{x+y}$ (C) xy

(D) $\frac{1}{xy}$ (E) $\frac{x+y}{xy}$

8. Two equal circles in the same plane cannot have the following number of common tangents:

(A) 1 (B) 2 (C) 3

(D) 4 (E) none of these

9. If $m = \frac{cab}{a-b}$, then b equals:

(A) $\frac{m(a-b)}{ca}$ (B) $\frac{cab-ma}{-m}$ (C) $\frac{1}{1+c}$

(D) $\frac{ma}{m+ca}$ (E) $\frac{m+ca}{ma}$

10. An automobile went up a hill at a speed of 10 miles an hour and down the same distance at a speed of 20 miles an hour. The average speed for the round trip was:

(A) $12\frac{1}{2}$ mph (B) $13\frac{1}{3}$ mph (C) $14\frac{1}{2}$ mph

(D) 15 mph (E) none of these

11. If $y = f(x) = \frac{x+2}{x-1}$, then it is incorrect to say:

(A) $x = \frac{y+2}{y-1}$ (B) $f(0) = -2$ (C) $f(1) = 0$

(D) $f(-2) = 0$ (E) $f(y) = x$

12. The sum to infinity of the terms of an infinite geometric progression is 6. The sum of the first two terms is $4\frac{1}{2}$. The first term of the progression is:

(A) 3 or $1\frac{1}{2}$ (B) 1 (C) $2\frac{1}{2}$

(D) 6 (E) 9 or 3

13. The function $x^2 + px + q$ with p and q greater than zero has its minimum value when:

(A) $x = -p$ (B) $x = \frac{p}{2}$ (C) $x = -2p$

(D) $x = \frac{p^2}{4q}$ (E) $x = \frac{-p}{2}$

14. A house and store were sold for \$12,000 each. The house was sold at a loss of 20% of the cost, and the store at a gain of 20% of the cost. The entire transaction resulted in:

(A) no loss or gain (B) loss of \$1000 (C) gain of \$1000
(D) gain of \$2000 (E) none of these

15. The sides of a triangle are in the ratio $6:8:9$. Then:

(A) the triangle is obtuse
(B) the angles are in the ratio $6:8:9$
(C) the triangle is acute
(D) the angle opposite the largest side is double the angle opposite the smallest side
(E) none of these

Part 2

16. If the base of a rectangle is increased by 10% and the area is unchanged, then the altitude is decreased by:

(A) 9% (B) 10% (C) 11%
(D) $11\frac{1}{9}\%$ (E) $9\frac{1}{11}\%$

17. A merchant bought some goods at a discount of 20% of the list price. He wants to mark them at such a price that he can give a discount of 20% of the marked price and still make a profit of 20% of the selling price. The per cent of the list price at which he should mark them is:

(A) 20 (B) 100 (C) 125
(D) 80 (E) 120

18. $\text{Log}\, p + \log q = \log(p+q)$ only if:

(A) $p = q =$ zero (B) $p = \dfrac{q^2}{1-q}$ (C) $p = q = 1$
(D) $p = \dfrac{q}{q-1}$ (E) $p = \dfrac{q}{q+1}$

19. Angle B of triangle ABC is trisected by BD and BE which meet AC at D and E respectively. Then:

(A) $\dfrac{\overline{AD}}{\overline{EC}} = \dfrac{\overline{AE}}{\overline{DC}}$ (B) $\dfrac{\overline{AD}}{\overline{EC}} = \dfrac{\overline{AB}}{\overline{BC}}$
(C) $\dfrac{\overline{AD}}{\overline{EC}} = \dfrac{\overline{BD}}{\overline{BE}}$ (D) $\dfrac{\overline{AD}}{\overline{EC}} = \dfrac{(\overline{AB})(\overline{BD})}{(\overline{BE})(\overline{BC})}$
(E) $\dfrac{\overline{AD}}{\overline{EC}} = \dfrac{(\overline{AE})(\overline{BD})}{(\overline{DC})(\overline{BE})}$

20. If $x/y = 3/4$, then the incorrect expression in the following is:

(A) $\frac{x+y}{y} = \frac{7}{4}$ (B) $\frac{y}{y-x} = \frac{4}{1}$

(C) $\frac{x+2y}{x} = \frac{11}{3}$ (D) $\frac{x}{2y} = \frac{3}{8}$

(E) $\frac{x-y}{y} = \frac{1}{4}$

21. The sides of a regular polygon of n sides, $n > 4$, are extended to form a star. The number of degrees at each point of the star is:

(A) $\frac{360}{n}$ (B) $\frac{(n-4)180}{n}$

(C) $\frac{(n-2)180}{n}$ (D) $180 - \frac{90}{n}$

(E) $\frac{180}{n}$

22. On hypotenuse AB of a right triangle ABC a second right triangle ABD is constructed with hypotenuse AB. If $\overline{BC} = 1$, $\overline{AC} = b$, and $\overline{AD} = 2$, then $\overline{BD}$ equals:

(A) $\sqrt{b^2+1}$ (B) $\sqrt{b^2-3}$ (C) $\sqrt{b^2+1}+2$

(D) b^2+5 (E) $\sqrt{b^2+3}$

23. If $\frac{x^2-bx}{ax-c} = \frac{m-1}{m+1}$ has roots which are numerically equal but of opposite signs, the value of m must be:

(A) $\frac{a-b}{a+b}$ (B) $\frac{a+b}{a-b}$ (C) c

(D) $\frac{1}{c}$ (E) 1

24. In the figure, it is given that angle $C = 90°$, $\overline{AD} = \overline{DB}$, $DE \perp AB$, $\overline{AB} = 20$, and $\overline{AC} = 12$. The area of quadrilateral $ADEC$ is:

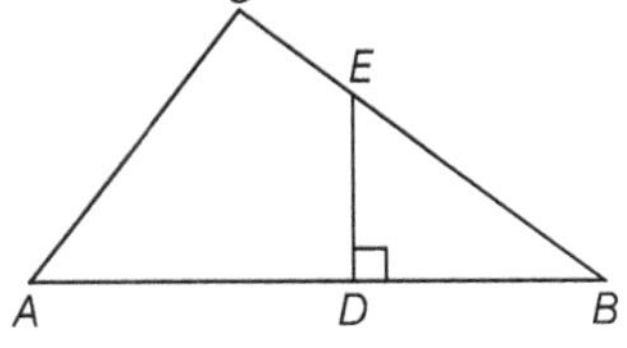

(A) 75 (B) $58\frac{1}{2}$ (C) 48

(D) $37\frac{1}{2}$ (E) none of these

25. A powderman set a fuse for a blast to take place is 30 seconds. He ran away at a rate of 8 yards per second. Sound travels at the rate of 1080 feet per second. When the powderman heard the blast, he had run approximately:

(A) 200 yd. (B) 352 yd. (C) 300 yd.
(D) 245 yd. (E) 512 yd.

26. If $\left(r + \frac{1}{r}\right)^2 = 3$, then $r^3 + \frac{1}{r^3}$ equals

(A) 1 (B) 2 (C) 0
(D) 3 (E) 6

27. The ratio of the perimeter of an equilateral triangle having an altitude equal to the radius of a circle, to the perimeter of an equilateral triangle inscribed in the circle is:

(A) $1:2$ (B) $1:3$ (C) $1:\sqrt{3}$
(D) $\sqrt{3}:2$ (E) $2:3$

28. In the table shown, the formula relating x and y is:

x	1	2	3	4	5
y	3	7	13	21	31

(A) $y = 4x - 1$
(B) $y = x^3 - x^2 + x + 2$
(C) $y = x^2 + x + 1$
(D) $y = (x^2 + x + 1)(x - 1)$
(E) none of these

29. In a circle of radius 5 units, CD and AB are perpendicular diameters. A chord CH cutting AB at K is 8 units long. The diameter AB is divided into two segments whose dimensions are:

(A) 1.25, 8.75 (B) 2.75, 7.25 (C) 2, 8
(D) 4, 6 (E) none of these

30. When the sum of the first ten terms of an arithmetic progression is four times the sum of the first five terms, the ratio of the first term to the common difference is:

(A) $1:2$ (B) $2:1$ (C) $1:4$
(D) $4:1$ (E) $1:1$

31. Given 12 points in a plane no three of which are collinear, the number of lines they determine is:

(A) 24 (B) 54 (C) 120
(D) 66 (E) none of these

32. K takes 30 minutes less time than M to travel a distance of 30 miles. K travels $\frac{1}{3}$ mile per hour faster than M. If x is K's rate of speed in miles per hour, then K's time for the distance is:

(A) $\frac{x+(1/3)}{30}$ (B) $\frac{x-(1/3)}{30}$ (C) $\frac{30}{x+(1/3)}$

(D) $\frac{30}{x}$ (E) $\frac{x}{30}$

33. A circle and a square have the same perimeter. Then:

(A) their areas are equal

(B) the area of the circle is the greater

(C) the area of the square is the greater

(D) the area of the circle is π times the area of the square

(E) none of these

34. The price of an article was increased $p\%$. Later the new price was decreased $p\%$. If the last price was one dollar, the original price was:

(A) $\frac{1-p^2}{200}$ (B) $\frac{\sqrt{1-p^2}}{100}$ (C) one dollar

(D) $1-\frac{p^2}{10{,}000-p^2}$ (E) $\frac{10{,}000}{10{,}000-p^2}$

35. With a rational denominator, the expression $\frac{\sqrt{2}}{\sqrt{2}+\sqrt{3}-\sqrt{5}}$ is equivalent to:

(A) $\frac{3+\sqrt{6}+\sqrt{15}}{6}$ (B) $\frac{\sqrt{6}-2+\sqrt{10}}{6}$

(C) $\frac{2+\sqrt{6}+\sqrt{10}}{10}$ (D) $\frac{2+\sqrt{6}-\sqrt{10}}{6}$

(E) none of these

Part 3

36. To be continuous at $x=-1$, the value of $\frac{x^3+1}{x^2-1}$ is taken to be:

(A) -2 (B) 0 (C) $\frac{3}{2}$

(D) ∞ (E) $-\frac{3}{2}$

37. Two equal parallel chords are drawn 8 inches apart in a circle of radius 8 inches. The area of that part of the circle that lies between the chords is:

(A) $21\frac{1}{3}\pi - 32\sqrt{3}$ (B) $32\sqrt{3} + 21\frac{1}{3}\pi$

(C) $32\sqrt{3} + 42\frac{2}{3}\pi$ (D) $16\sqrt{3} + 42\frac{2}{3}\pi$

(E) $42\frac{2}{3}\pi$

38. The area of a trapezoidal field is 1400 square yards. Its altitude is 50 yards. Find the two bases, if the number of yards in each base is an integer divisible by 8. The number of solutions to this problem is:

(A) none (B) one (C) two

(D) three (E) more than three

39. If the perimeter of a rectangle is p and its diagonal is d, the difference between the length and width of the rectangle is:

(A) $\frac{\sqrt{8d^2 - p^2}}{2}$ (B) $\frac{\sqrt{8d^2 + p^2}}{2}$ (C) $\frac{\sqrt{6d^2 - p^2}}{2}$

(D) $\frac{\sqrt{6d^2 + p^2}}{2}$ (E) $\frac{\sqrt{8d^2 - p^2}}{4}$

40. In order to draw a graph of $f(x) = ax^2 + bx + c$, a table of values was constructed. These values of the function for a set of equally spaced increasing values of x were 3844, 3969, 4096, 4227, 4356, 4489, 4624, and 4761. The one which is incorrect is:

(A) 4096 (B) 4356 (C) 4489

(D) 4761 (E) none of these

41. Increasing the radius of a cylinder by 6 units increases the volume by y cubic units. Increasing the altitude of the cylinder by 6 units also increases the volume by y cubic units. If the original altitude is 2, then the original radius is:

(A) 2 (B) 4 (C) 6

(D) 6π (E) 8

42. Let D represent a repeating decimal. If P denotes the r figures of D which do not repeat themselves, and Q denotes the s figures which do repeat themselves, then the incorrect expression is:

(A) $D = .PQQQ\ldots$ (B) $10^r D = P.QQQ\ldots$

(C) $10^{r+s} D = PQ.QQQ\ldots$ (D) $10^r(10^s - 1)D = Q(P - 1)$

(E) $10^r.10^{2s} D = PQQ.QQQ\ldots$

43. The diameter of a circle is divided into n equal parts. On each part a semi circle is constructed. As n becomes very large, the sum of the lengths of the arcs of the semicircles approaches a length:

(A) equal to the semi-circumference of the original circle

(B) equal to the diameter of the original circle

(C) greater than the diameter but less than the semi-circumference of the original circle

(D) that is infinite

(E) greater than the semi-circumference but finite

44. If an integer of two digits is k times the sum of its digits, the number formed by interchanging the digits is the sum of the digits multiplied by:

(A) $(9 - k)$ (B) $(10 - k)$ (C) $(11 - k)$

(D) $(k - 1)$ (E) $(k + 1)$

45. If a and b are two unequal positive numbers, then:

(A) $\dfrac{2ab}{a+b} > \sqrt{ab} > \dfrac{a+b}{2}$ (B) $\sqrt{ab} > \dfrac{2ab}{a+b} > \dfrac{a+b}{2}$

(C) $\dfrac{2ab}{a+b} > \dfrac{a+b}{2} > \sqrt{ab}$ (D) $\dfrac{a+b}{2} > \dfrac{2ab}{a+b} > \sqrt{ab}$

(E) $\dfrac{a+b}{2} > \sqrt{ab} > \dfrac{2ab}{a+b}$

46. The base of a new rectangle equals the sum of the diagonal and the greater side of a given rectangle, while the altitude of the new rectangle equals the difference of the diagonal and the greater side of the given rectangle.

The area of the new rectangle is:

(A) greater than the area of the given rectangle

(B) equal to the area of the given rectangle

(C) equal to the area of a square with its side equal to the smaller side of the given rectangle

(D) equal to the area of a square with its side equal to the greater side of the given rectangle

(E) equal to the area of a rectangle whose dimensions are the diagonal and shorter side of the given rectangle

47. In the set of equations $z^x = y^{2x}$, $2^z = 2.4^x$, $x + y + z = 16$, the integral roots in the order x, y, z are:

(A) 3, 4, 9 (B) 9, – 5, 12 (C) 12, – 5, 9

(D) 4, 3, 9 (E) 4, 9, 3

48. Two cyclists, k miles apart, and starting at the same time, would be together in r hours if they traveled in the same direction, but would pass each other in t hours if they traveled in opposite directions. The ratio of the speed of the faster cyclist to that of the slower is:

(A) $\frac{r+t}{r-t}$ (B) $\frac{r}{r-t}$ (C) $\frac{r+t}{r}$

(D) $\frac{r}{t}$ (E) $\frac{r+k}{t-k}$

49. In the figure, $\overline{CD}$, $\overline{AE}$ and $\overline{BF}$ are one-third of their respective sides. It follows that $\overline{AN_2}:\overline{N_2N_1}:\overline{N_1D}=3:3:1$, and similarly for lines BE and CF. Then, the area of triangle $N_1N_2N_3$ is:

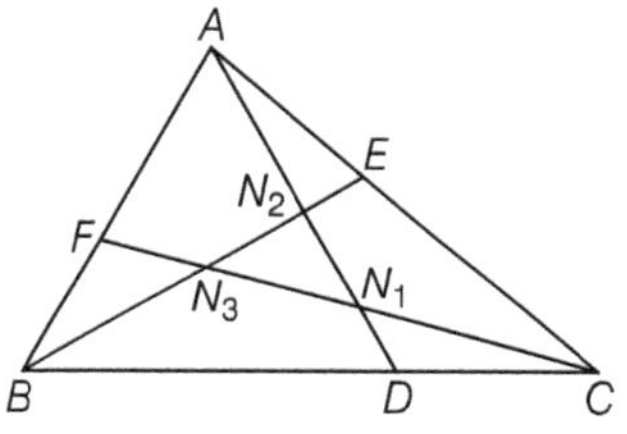

(A) $\frac{1}{10}\,\Delta ABC$ (B) $\frac{1}{9}\,\Delta ABC$ (C) $\frac{1}{7}\,\Delta ABC$

(D) $\frac{1}{6}\,\Delta ABC$ (E) none of these

50. A line initially 1 inch long grows according to the following law, where the first term is the initial length.

$1+\frac{1}{4}\sqrt{2}+\frac{1}{4}+\frac{1}{16}\sqrt{2}+\frac{1}{16}+\frac{1}{64}\sqrt{2}+\frac{1}{64}+\dots$.

If the growth process continues forever, the limit of the length of the line is:

(A) ∞ (B) $\frac{4}{3}$ (C) $\frac{8}{3}$

(D) $\frac{1}{3}(4+\sqrt{2})$ (E) $\frac{2}{3}(4+\sqrt{2})$

1953 Examination

Part 1

1. A boy buys oranges at 3 for 10 cents. He will sell them at 5 for 20 cents. In order to make a profit of \$1.00, he must sell:

(A) 67 oranges
(B) 150 oranges
(C) 200 oranges
(D) an infinite number of oranges
(E) none of these

2. A refrigerator is offered for sale at \$250.00 less successive discounts of 20% and 15%. The sale price of the refrigerator is:

(A) 35% less than \$250.00
(B) 65% of \$250.00
(C) 77% of \$250.00
(D) 68% of \$250.00
(E) none of these

3. The factors of the expression $x^2 + y^2$ are:

(A) $(x + y)(x - y)$
(B) $(x + y)^2$
(C) $(x^{2/3} + y^{2/3})(x^{4/3} + y^{4/3})$
(D) $(x + iy)(x - iy)$
(E) none of these

4. The roots of $x(x^2 + 8x + 16)(4 - x) = 0$ are:

(A) 0
(B) 0, 4
(C) 0, 4, −4
(D) 0, 4, −4, −4
(E) none of these

5. If $\log_6 x = 2.5$, the value of x is:

(A) 90
(B) 36
(C) $36\sqrt{6}$
(D) 0.5
(E) none of these

6. Charles has $5q + 1$ quarters and Richard has $q + 5$ quarters. The difference in their money in dimes is:

(A) $10(q - 1)$
(B) $\frac{2}{5}(4q - 4)$
(C) $\frac{2}{5}(q - 1)$
(D) $\frac{5}{2}(q - 1)$
(E) none of these

7. The fraction $\dfrac{\sqrt{a^2+x^2}-(x^2-a^2)/\sqrt{a^2+x^2}}{a^2+x^2}$ reduces to:

(A) 0 (B) $\dfrac{2a^2}{(a^2+x^2)}$ (C) $\dfrac{2x^2}{(a^2+x^2)^{3/2}}$

(D) $\dfrac{2a^2}{(a^2+x^2)^{3/2}}$ (E) $\dfrac{2x^2}{a^2+x^2}$

8. The value of x at the intersection of $y=8/(x^2+4)$ and $x+y=2$ is:

(A) $-2+\sqrt{5}$ (B) $-2-\sqrt{5}$ (C) 0

(D) 2 (E) none of these

9. The number of ounces of water needed to reduce 9 ounces of shaving lotion containing 50% alcohol to a lotion containing 30% alcohol is:

(A) 3 (B) 4 (C) 5

(D) 6 (E) 7

10. The number of revolutions of a wheel, with fixed centre and with an outside diameter of 6 feet, required to cause a point on the rim to go one mile is:

(A) 880 (B) $\dfrac{440}{\pi}$ (C) $\dfrac{880}{\pi}$

(D) 440π (E) none of these

11. A running track is the ring formed by two concentric circles. It is 10 feet wide. The circumferences of the two circles differ by about:

(A) 10 feet (B) 30 feet (C) 60 feet

(D) 100 feet (E) none of these

12. The diameters of two circles are 8 inches and 12 inches, respectively. The ratio of the area of the smaller to the area of the larger circle is:

(A) $\dfrac{2}{3}$ (B) $\dfrac{4}{9}$ (C) $\dfrac{9}{4}$

(D) $\dfrac{1}{2}$ (E) none of these

13. A triangle and a trapezoid are equal in area. They also have the same altitude. If the base of the triangle is 18 inches, the median of the trapezoid is:

(A) 36 inches

(B) 9 inches

(C) 18 inches

(D) not obtainable from these data

(E) none of these

14. Given the larger of two circles with centre P and radius p and the smaller with centre Q and radius q. Draw PQ. Which of the following statements is false?

(A) $p - q$ can be equal to $\overline{PQ}$ (B) $p + q$ can be equal to $\overline{PQ}$
(C) $p + q$ can be less than $\overline{PQ}$ (D) $p - q$ can be less than $\overline{PQ}$
(E) none of these

15. A circular piece of metal of maximum size is cut out of a square piece and then a square piece of maximum size is cut out of the circular piece. The total amount of metal wasted is:

(A) $\frac{1}{4}$ the area of the original square

(B) $\frac{1}{2}$ the area of the original square

(C) $\frac{1}{2}$ the area of the circular piece

(D) $\frac{1}{4}$ the area of the circular piece

(E) none of these

Part 2

16. Adams plans a profit of 10% on the selling price of an article and his expenses are 15% of sales. The rate of mark-up on an article that sells for \$5.00 is:

(A) 20% (B) 25% (C) 30%
(D) $33\frac{1}{3}\%$ (E) 35%

17. A man has part of \$4500 invested at 4% and the rest at 6%. If his annual return on each investment is the same, the average rate of interest which he realizes on the \$4500 is:

(A) 5% (B) 4.8% (C) 5.2%
(D) 4.6% (E) none of these

18. One of the factors of $x^4 + 4$ is:

(A) $x^2 + 2$ (B) $x + 1$ (C) $x^2 - 2x + 2$
(D) $x^2 - 4$ (E) none of these

19. In the expression xy^2, the values of x and y are each decreased 25%, the value of the expression is:

(A) decreased 50% (B) decreased 75%
(C) decreased 37/64 of its value (D) decreased 27/64 of its value
(E) none of these

20. If $y = x + \frac{1}{x}$, then $x^4 + x^3 - 4x^2 + x + 1 = 0$ becomes:

(A) $x^2(y^2 + y - 2) = 0$ (B) $x^2(y^2 + y - 3) = 0$
(C) $x^2(y^2 + y - 4) = 0$ (D) $x^2(y^2 + y - 6) = 0$
(E) none of these

21. If $\log_{10}(x^2 - 3x + 6) = 1$, the value of x is:

(A) 10 or 2 (B) 4 or -2 (C) 3 or -1
(D) 4 or -1 (E) none of these

22. The logarithm of $27\sqrt[4]{9}\,\sqrt[3]{9}$ to the base 3 is:

(A) $8\frac{1}{2}$ (B) $4\frac{1}{6}$ (C) 5
(D) 3 (E) none of these

23. The equation $\sqrt{x+10} - \frac{6}{\sqrt{x+10}} = 5$ has:

(A) an extraneous root between -5 and -1
(B) an extraneous root between -10 and -6
(C) a true root between 20 and 25
(D) two true roots
(E) two extraneous roots

24. If a, b and c are positive integers less than 10, then $(10a + b)(10a + c)$ equals $100a(a + 1) + bc$ if:

(A) $b + c = 10$ (B) $b = c$ (C) $a + b = 10$
(D) $a = b$ (E) $a + b + c = 10$

25. In a geometric progression whose terms are positive, any term is equal to the sum of the next two following terms. Then the common ratio is:

(A) 1 (B) about $\frac{\sqrt{5}}{2}$ (C) $\frac{\sqrt{5}-1}{2}$
(D) $\frac{1-\sqrt{5}}{2}$ (E) $\frac{2}{\sqrt{5}}$

26. The base of a triangle is 15 inches. Two lines are drawn parallel to the base, terminating in the other two sides, and dividing the triangle into three equal areas. The length of the parallel closer to the base is:

(A) $5\sqrt{6}$ inches (B) 10 inches (C) $4\sqrt{3}$ inches
(D) 7.5 inches (E) none of these

27. The radius of the first circle is 1 inch, that of the second $\frac{1}{2}$ inch, that of the third $\frac{1}{4}$ inch and so on indefinitely. The sum of the areas of the circles is:

(A) $\frac{3\pi}{4}$ (B) 1.3π (C) 2π

(D) $\frac{4\pi}{3}$ (E) none of these

28. In triangle ABC, sides a, b and c are opposite angles A, B and C respectively. AD bisects angle A and meets BC at D. Then if $x = \overline{CD}$ and $y = \overline{BD}$ the correct proportion is:

(A) $x/a = a/(b+c)$ (B) $x/b = a/(a+c)$ (C) $y/c = c/(b+c)$

(D) $y/c = a/(b+c)$ (E) $x/y = c/b$

29. The number of significant digits in the measurement of the side of a square whose computed area is 1.1025 square inches to the nearest ten-thousandth of a square inch is:

(A) 2 (B) 3 (C) 4

(D) 5 (E) 1

30. A house worth \$9000 is sold by Mr. A to Mr. B at a 10% loss. Mr. B sells the house back to Mr. A at a 10% gain. The result of the two transactions is:

(A) Mr. A breaks even

(B) Mr. B gains \$900

(C) Mr. A loses \$900

(D) Mr. A loses \$810

(E) Mr. B gains \$1710

31. The rails on a railroad are 30 feet long. As the train passes over the point where the rails are joined, there is an audible click. The speed of the train in miles per hour is approximately the number of clicks heard in:

(A) 20 seconds (B) 2 minutes (C) $1\frac{1}{2}$ minutes

(D) 5 minutes (E) none of these

32. Each angle of a rectangle is trisected. The intersections of the pairs of trisectors adjacent to the same side always form:

(A) a square (B) a rectangle

(C) a parallelogram with unequal sides

(D) a rhombus

(E) a quadrilateral with no special properties

33. The perimeter of an isosceles right triangle is $2p$. Its area is:

(A) $(2+\sqrt{2})p$ (B) $(2-\sqrt{2})p$ (C) $(3-2\sqrt{2})p^2$
(D) $(1-2\sqrt{2})p^2$ (E) $(3+2\sqrt{2})p^2$

34. If one side of a triangle is 12 inches and the opposite angles is 30 degrees, then the diameter of the circumscribed circle is:

(A) 18 inches (B) 30 inches (C) 24 inches
(D) 20 inches (E) none of these

35. If $f(x) = \dfrac{x(x-1)}{2}$, then $f(x+2)$ equals:

(A) $f(x) + f(2)$ (B) $(x+2)\,f(x)$ (C) $x(x+2)\,f(x)$
(D) $\dfrac{x\,f(x)}{x+2}$ (E) $\dfrac{(x+2)\,f(x+1)}{x}$

Part 3

36. Determine m so that $4x^2 - 6x + m$ is divisible by $x - 3$. The obtained value, m, is an exact divisor of:

(A) 12 (B) 20 (C) 36
(D) 48 (E) 64

37. The base of an isosceles triangle is 6 inches and one of the equal sides is 12 inches. The radius of the circle through the vertices of the triangle is:

(A) $\dfrac{7\sqrt{15}}{5}$ (B) $4\sqrt{3}$ (C) $3\sqrt{5}$
(D) $6\sqrt{3}$ (E) none of these

38. If $f(a) = a - 2$ and $F(a, b) = b^2 + a$, then $F[3, f(4)]$ is:

(A) $a^2 - 4a + 7$ (B) 28 (C) 7
(D) 8 (E) 11

39. The product, $\log_a b \cdot \log_b a$ is equal to:

(A) 1 (B) a (C) b
(D) ab (E) none of these

40. The negation of the statement "all men are honest," is:

(A) no men are honest
(B) all men are dishonest
(C) some men are dishonest
(D) no men are dishonest
(E) some men are honest

41. A girls' camp is located 300 rods from a straight road. On this road, a boys' camp is located 500 rods from the girls' camp. It is desired to build a canteen on the road which shall be exactly the same distance from each camp. The distance of the canteen from each of the camps is:

(A) 400 rods (B) 250 rods (C) 87.5 rods
(D) 200 rods (E) none of these

42. The centers of two circles are 41 inches apart. The smaller circle has a radius of 4 inches and the larger one has a radius of 5 inches. The length of the common internal tangent is:

(A) 41 inches (B) 39 inches (C) 39.8 inches
(D) 40.1 inches (E) 40 inches

43. If the price of an article is increased by per cent p, then the decrease in per cent of sales must not exceed d in order to yield the same income. The value of d is:

(A) $\frac{1}{1+p}$ (B) $\frac{1}{1-p}$ (C) $\frac{p}{1+p}$
(D) $\frac{p}{p-1}$ (E) $\frac{1-p}{1+p}$

44. In solving a problem that reduces to a quadratic equation one student makes a mistake only in the constant term of the equation and obtains 8 and 2 for the roots. Another student makes a mistake only in the coefficient of the first degree term and finds -9 and -1 for the roots. The correct equation was:

(A) $x^2-10x+9=0$ (B) $x^2+10x+9=0$ (C) $x^2-10x+16=0$
(D) $x^2-8x-9=0$ (E) none of these

45. The lengths of two line segments are a units and b units, respectively. Then, the correct relation between them is:

(A) $\frac{(a+b)}{2}>\sqrt{ab}$ (B) $\frac{(a+b)}{2}<\sqrt{ab}$ (C) $\frac{(a+b)}{2}=\sqrt{ab}$
(D) $\frac{(a+b)}{2}\leqq\sqrt{ab}$ (E) $\frac{(a+b)}{2}\geqq\sqrt{ab}$

46. Instead of walking along two adjacent sides of a rectangular field, a boy took a short-cut along the diagonal of the field and saved a distance equal to 1/2 the longer side. The ratio of the shorter side of the rectangle to the longer side was:

(A) $\frac{1}{2}$ (B) $\frac{2}{3}$ (C) $\frac{1}{4}$
(D) $\frac{3}{4}$ (E) $\frac{2}{5}$

47. If x is greater than zero, then the correct relationship is:

(A) $\log(1 + x) = x/(1 + x)$

(B) $\log(1 + x) < x/(1 + x)$

(C) $\log(1 + x) > x$

(D) $\log(1 + x) < x$

(E) none of these

48. If the larger base of an isosceles trapezoid equals a diagonal and the smaller base equals the altitude, then the ratio of the smaller base to the larger base is:

(A) 1/2 (B) 2/3 (C) 3/4
(D) 3/5 (E) 2/5

49. The coordinates of A, B and C are (5, 5), (2, 1) and (0, k) respectively. The value of k that makes $\overline{AC} + \overline{BC}$ as small as possible is:

(A) 3 (B) $4\frac{1}{2}$ (C) $3\frac{6}{7}$
(D) $4\frac{5}{6}$ (E) $2\frac{1}{7}$

50. One of the sides of a triangle is divided into segments of 6 and 8 units by the point of tangency of the inscribed circle. If the radius of the circle is 4, then the length of the shortest side of the triangle is:

(A) 12 units (B) 13 units (C) 14 units
(D) 15 units (E) 16 units

1954 Examination

Part 1

1. The square of $5-\sqrt{y^2-25}$ is:

(A) $y^2-5\sqrt{y^2-25}$ (B) $-y^2$ (C) y^2

(D) $(5-y)^2$ (E) $y^2-10\sqrt{y^2-25}$

2. The equation $\frac{2x^2}{x-1}-\frac{2x+7}{3}+\frac{4-6x}{x-1}+1=0$ can be transformed by eliminating fractions to the equation $x^2-5x+4=0$. The roots of the latter equation are 4 and 1. Then the roots of the first equation are:

(A) 4 and 1 (B) only 1 (C) only 4

(D) neither 4 nor 1 (E) 4 and some other root

3. If x varies as the cube of y, and y varies as the fifth root of z, then x varies as the nth power of z, where n is:

(A) $\frac{1}{15}$ (B) $\frac{5}{3}$ (C) $\frac{3}{5}$

(D) 15 (E) 8

4. If the Highest Common Divisor of 6432 and 132 is diminished by 8, it will equal:

(A) -6 (B) 6 (C) -2

(D) 3 (E) 4

5. A regular hexagon is inscribed in a circle of radius 10 inches. Its area is:

(A) $150\sqrt{3}$ sq.in. (B) 150 sq.in. (C) $25\sqrt{3}$ sq.in.

(D) 600 sq.in. (E) $300\sqrt{3}$ sq.in.

6. The value of $\left(\frac{1}{16}\right)a^0+\left(\frac{1}{16a}\right)^0-64^{-1/2}-(-32)^{-4/5}$ is:

(A) $1\frac{13}{16}$ (B) $1\frac{3}{16}$ (C) 1

(D) $\frac{7}{8}$ (E) $\frac{1}{16}$

7. A housewife saved \$2.50 in buying a dress on sale. If she spent \$25 for the dress, she saved about:

(A) 8% (B) 9% (C) 10%
(D) 11% (E) 12%

8. The base of a triangle is twice as long as a side of a square and their areas are the same. Then the ratio of the altitude of the triangle to the side of the square is:

(A) $\frac{1}{4}$ (B) $\frac{1}{2}$ (C) 1
(D) 2 (E) 4

9. A point P is outside a circle and is 13 inches from the center. A secant from P cuts the circle at Q and R so that the external segment of the secant PQ is 9 inches and $\overline{QR}$ is 7 inches. The radius of the circle is:

(A) 3″ (B) 4″ (C) 5″
(D) 6″ (E) 7″

10. The sum of the numerical coefficients in the expansion of the binomial $(a + b)^6$ is:

(A) 32 (B) 16 (C) 64
(D) 48 (E) 7

11. A merchant placed on display some dresses, each with a marked price. He then posted a sign "$\frac{1}{3}$ off on these dresses.". The cost of the dresses was $\frac{3}{4}$ of the price at which he actually sold them. Then the ratio of the cost to the marked price was:

(A) $\frac{1}{2}$ (B) $\frac{1}{3}$ (C) $\frac{1}{4}$
(D) $\frac{2}{3}$ (E) $\frac{3}{4}$

12. The solution of the equations $\begin{matrix} 2x-3y=7 \\ 4x-6y=20 \end{matrix}$ is:

(A) $x=18, y=12$ (B) $x=0, y=0$ (C) there is no solution
(D) there is an unlimited number of solutions
(E) $\begin{matrix} x=8 \\ y=5 \end{matrix}$

13. A quadrilateral is inscribed in a circle. If angles are inscribed in the four arcs cut off by the sides of the quadrilateral, without intersecting the sides between vertices, their sum will be:

(A) 180° (B) 540° (C) 360°
(D) 450° (E) 1080°

14. When simplified $\sqrt{1+\left(\frac{x^4-1}{2x^2}\right)^2}$ equals:

(A) $\frac{x^4+2x^2-1}{2x^2}$ (B) $\frac{x^4-1}{2x^2}$ (C) $\frac{\sqrt{x^2+1}}{2}$

(D) $\frac{x^2}{\sqrt{2}}$ (E) $\frac{x^2}{2}+\frac{1}{2x^2}$

15. Log 125 equals:

(A) 100 log 1.25 (B) 5 log 3 (C) 3 log 25

(D) 3 – 3 log 2 (E) (log 25) (log 5)

Part 2

16. If $f(x) = 5x^2 - 2x - 1$, then $f(x+h) - f(x)$ equals:

(A) $5h^2 - 2h$ (B) $10xh - 4x + 2$ (C) $10xh - 2x - 2$

(D) $h(10x + 5h - 2)$ (E) $3h$

17. The graph of the function $f(x) = 2x^3 - 7$ goes:

(A) up to the right and down to the left

(B) down to the right and up to the left

(C) up to the right and up to the left

(D) down to the right and down to the left

(E) none of these ways.

18. Of the following sets, the one that includes all values of x which will satisfy $2x - 3 > 7 - x$ is:

(A) $x > 4$ (B) $x < \frac{10}{3}$ (C) $x = \frac{10}{3}$

(D) $x > \frac{10}{3}$ (E) $x < 0$

19. If the three points of contact of a circle inscribed in a triangle are joined, the angles of the resulting triangle:

(A) are always equal to 60°

(B) are always one obtuse angle and two unequal acute angles

(C) are always one obtuse angle and two equal acute angles

(D) are always acute angles

(E) are always unequal to each other

20. The equation $x^3 + 6x^2 + 11x + 6 = 0$ has:

(A) no negative real roots (B) no positive real roots

(C) no real roots (D) 1 positive and 2 negative roots

(E) 1 negative and 2 positive roots

21. The roots of the equation $2\sqrt{x} + 2x^{-1/2} = 5$ can be found by solving:

(A) $16x^2 - 92x + 1 = 0$ (B) $4x^2 - 25x + 4 = 0$
(C) $4x^2 - 17x + 4 = 0$ (D) $2x^2 - 21x + 2 = 0$
(E) $4x^2 - 25x - 4 = 0$

22. The expression $\frac{2x^2 - x}{(x+1)(x-2)} - \frac{4+x}{(x+1)(x-2)}$ cannot be evaluated for $x = -1$ or $x = 2$, since division by zero is not allowed. For other values of x:

(A) the expression takes on many different values
(B) the expression has only the value 2
(C) the expression has only the value 1
(D) the expression always has a value between -1 and $+2$
(E) the expression has a value greater than 2 or less than -1

23. If the margin made on an article costing C dollars and selling for S dollars is $M = \frac{1}{n}C$, then the margin is given by:

(A) $M = \frac{1}{n-1}S$ (B) $M = \frac{1}{n}S$ (C) $M = \frac{n}{n+1}S$
(D) $M = \frac{1}{n+1}S$ (E) $M = \frac{n}{n-1}S$

24. The values of k for which the equation $2x^2 - kx + x + 8 = 0$ will have real and equal roots are:

(A) 9 and -7 (B) only -7 (C) 9 and 7
(D) -9 and -7 (E) only 9

25. The two roots of the equation $a(b-c)x^2 + b(c-a)x + c(a-b) = 0$ are 1 and:

(A) $\frac{b(c-a)}{a(b-c)}$ (B) $\frac{a(b-c)}{c(a-b)}$ (C) $\frac{a(b-c)}{b(c-a)}$
(D) $\frac{c(a-b)}{a(b-c)}$ (E) $\frac{c(a-b)}{b(c-a)}$

26. The straight line AB is divided at C so that $\overline{AC} = 3\overline{CB}$. Circles are described on AC and CB as diameters and a common tangent meets AB produced at D. Then $\overline{BD}$ equals:

(A) the diameter of the smaller circle
(B) the radius of the smaller circle
(C) the radius of the larger circle
(D) $\overline{CB}\sqrt{3}$
(E) the difference of the two radii

27. A right circular cone has for its base a circle having the same radius as a given sphere. The volume of the cone is one-half that of the sphere. The ratio of the altitude of the cone to the radius of its base is:

(A) $\frac{1}{1}$ (B) $\frac{1}{2}$ (C) $\frac{2}{3}$

(D) $\frac{2}{1}$ (E) $\sqrt{\frac{5}{4}}$

28. If $\frac{m}{n}=\frac{4}{3}$ and $\frac{r}{t}=\frac{9}{14}$, the value of $\frac{3mr-nt}{4nt-7mr}$ is:

(A) $-5\frac{1}{2}$ (B) $-\frac{11}{14}$ (C) $-1\frac{1}{4}$

(D) $\frac{11}{14}$ (E) $-\frac{2}{3}$

29. If the ratio of the legs of a right triangle is 1 : 2, then the ratio of the corresponding segments of the hypotenuse made by a perpendicular upon it from the vertex is:

(A) 1 : 4 (B) $1:\sqrt{2}$ (C) 1 : 2

(D) $1:\sqrt{5}$ (E) 1 : 5

30. A and B together can do a job in 2 days; B and C can do it in four days; and A and C in $2\frac{2}{5}$ days. The number of days required for A to do the job alone is:

(A) 1 (B) 3 (C) 6

(D) 12 (E) 2.8

31. In triangle ABC, $\overline{AB}=\overline{AC}$, angle $A=40°$. Point O is within the triangle with angle OBC = angle OCA. The number of degrees in angle BOC is:

(A) 110 (B) 35 (C) 140

(D) 55 (E) 70

32. The factors of x^4+64 are:

(A) $(x^2+8)^2$

(B) $(x^2+8)(x^2-8)$

(C) $(x^2+2x+4)(x^2-8x+16)$

(D) $(x^2-4x+8)(x^2-4x-8)$

(E) $(x^2-4x+8)(x^2+4x+8)$

33. A bank charges $6 for a loan of $120. The borrower receives $114 and repays the loan in 12 installments of $10 a month. The interest rate is approximately:

(A) 5% (B) 6% (C) 7%

(D) 9% (E) 15%

34. The fraction, 1/3:

(A) equals 0.33333333

(B) is less than 0.33333333 by $\dfrac{1}{3\cdot 10^8}$

(C) is less than 0.33333333 by $\dfrac{1}{3\cdot 10^9}$

(D) is grater than 0.33333333 by $\dfrac{1}{3\cdot 10^8}$

(E) is greater than 0.33333333 by $\dfrac{1}{3\cdot 10^9}$

35. In the right triangle shown the sum of the distances $\overline{BM}$ and $\overline{MA}$ is equal to the sum of the distances $\overline{BC}$ and $\overline{CA}$. If $\overline{MB} = x$, $\overline{CB} = h$, and $\overline{CA} = d$, then x equals:

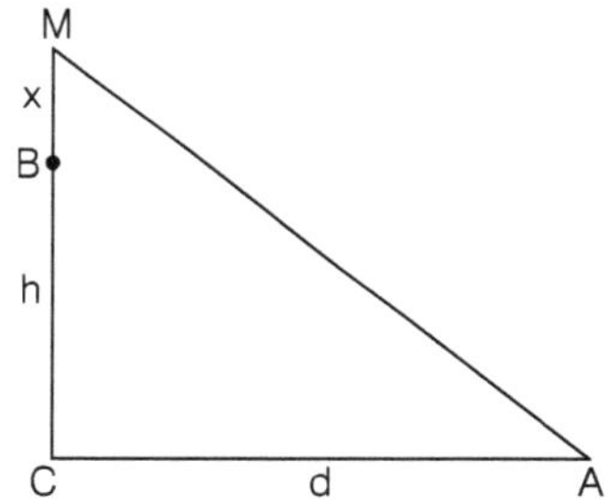

(A) $\dfrac{hd}{2h+d}$ (B) $d-h$

(C) $\dfrac{1}{2}d$ (D) $h+d-\sqrt{2d}$

(E) $\sqrt{h^2+d^2}-h$

Part 3

36. A boat has a speed of 15 mph in still water. In a stream that has a current of 5 mph it travels a certain distance downstream and returns. The ratio of the average speed for the round trip to the speed in still water is:

(A) $\dfrac{5}{4}$ (B) $\dfrac{1}{1}$ (C) $\dfrac{8}{9}$

(D) $\dfrac{7}{8}$ (E) $\dfrac{9}{8}$

37. Given triangle PQR with RS bisecting angle R, PQ extended to D and angle n a right angle, then:

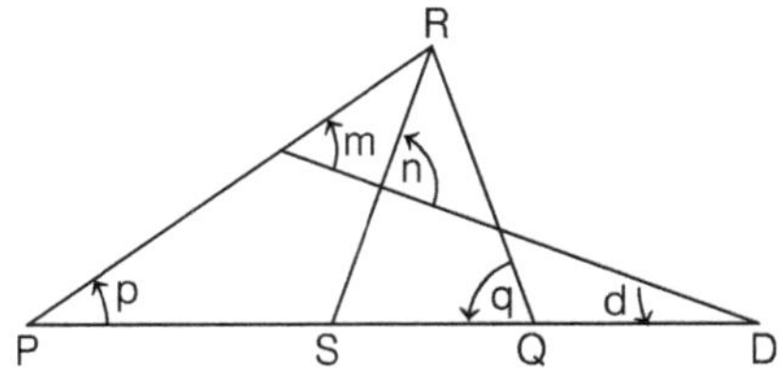

(A) $\measuredangle m = \frac{1}{2}(\measuredangle p - \measuredangle q)$
(B) $\measuredangle m = \frac{1}{2}(\measuredangle p + \quad q)$
(C) $\measuredangle d = \frac{1}{2}(\measuredangle q + \measuredangle p)$
(D) $\measuredangle d = \frac{1}{2}\measuredangle m$
(E) none of these is correct

38. If $\log 2 = .3010$ and $\log 3 = .4771$, the value of x when $3^{x+3} = 135$ is approximately:
(A) 5
(B) 1.47
(C) 1.67
(D) 1.78
(E) 1.63

39. The locus of the mid-point of a line segment that is drawn from a given external point P to a given circle with center O and radius r, is:
(A) a straight line perpendicular to PO
(B) a straight line parallel to PO
(C) a circle with centre P and radius r
(D) a circle with centre at the mid-point of PO and radius $2r$
(E) a circle with centre at the mid-point of PO and radius $\frac{1}{2}r$.

40. If $\left(a + \frac{1}{a}\right)^2 = 3$, then $a^3 + \frac{1}{a^3}$ equals:
(A) $\frac{10\sqrt{3}}{3}$
(B) $3\sqrt{3}$
(C) 0
(D) $7\sqrt{7}$
(E) $6\sqrt{3}$

41. The sum of all the roots of $4x^3 - 8x^2 - 63x - 9 = 0$ is:
(A) 8
(B) 2
(C) -8
(D) -2
(E) 0

42. Consider the graphs of (1)$y = x^2 - \frac{1}{2}x + 2$ and (2)$y = x^2 + \frac{1}{2}x + 2$ on the same set of axes. These parabolas have exactly the same shape. Then:

(A) the graphs coincide
(B) the graph of (1) is lower than the graph of (2)
(C) the graph of (1) is to the left of the graph of (2)
(D) the graph of (1) is to the right of the graph of (2)
(E) the graph of (1) is higher than the graph of (2)

43. The hypotenuse of a right triangle is 10 inches and the radius of the inscribed circle is 1 inch. The perimeter of the triangle in inches is:

(A) 15 (B) 22 (C) 24
(D) 26 (E) 30

44. A man born in the first half of the nineteenth century was x years old in the year x^2. He was born in:

(A) 1849 (B) 1825 (C) 1812
(D) 1836 (E) 1806

45. In a rhombus $ABCD$ line segments are drawn within the rhombus, parallel to diagonal BD, and terminated in the sides of the rhombus. A graph is drawn showing the length of a segment as a function of its distance from vertex A. The graph is:

(A) a straight line passing through the origin
(B) a straight line cutting across the upper right quadrant
(C) two line segments forming an upright V
(D) two line segments forming an inverted V, (Λ)
(E) none of these

46. In the diagram if points A, B, C are points of tangency, then x equals:

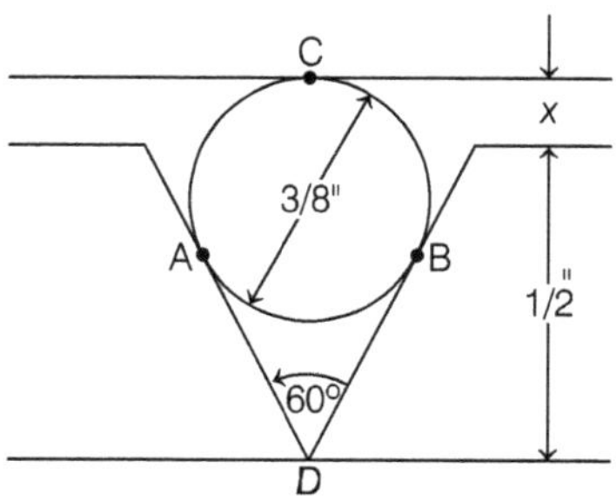

(A) $\frac{3''}{16}$ (B) $\frac{1''}{8}$ (C) $\frac{1''}{32}$
(D) $\frac{3''}{32}$ (E) $\frac{1''}{16}$

47. At the midpoint of line segment AB which is p units long, a perpendicular MR is erected with length q units. An arc is described from R with a radius equal to $\frac{1}{2}\overline{AB}$, meeting $\overline{AB}$ at T. Then, $\overline{AT}$ and $\overline{TB}$ are the roots of:

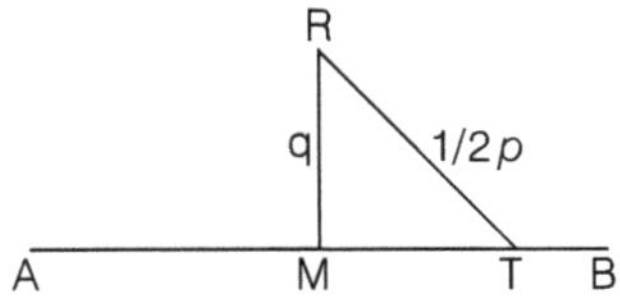

(A) $x^2 + px + q^2 = 0$ (B) $x^2 - px + q^2 = 0$
(C) $x^2 + px - q^2 = 0$ (D) $x^2 - px - q^2 = 0$
(E) $x^2 - px + q = 0$

48. A train, an hour after starting, meets with an accident which detains it a half hour, after which it proceeds at $\frac{3}{4}$ of its former rate and arrives $3\frac{1}{2}$ hours late. Had the accident happened 90 miles farther along the line, it would have arrived only 3 hours late. The length of the trip in miles was:

(A) 400 (B) 465 (C) 600
(D) 640 (E) 550

49. The difference of the squares of two odd numbers is always divisible by 8. If $a > b$, and $2a + 1$ and $2b + 1$ are the odd numbers, to prove the given statement we put the difference of the squares in the form:

(A) $(2a + 1)^2 - (2b + 1)^2$ (B) $4a^2 - 4b^2 + 4a - 4b$
(C) $4[a(a + 1) - b(b + 1)]$ (D) $4(a - b)(a + b + 1)$
(E) $4(a^2 + a - b^2 - b)$

50. The times between 7 and 8 o'clock, correct to the nearest minute, when the hands of a clock will form an angle of 84 degrees are:

(A) 7 : 23 and 7 : 53 (B) 7 : 20 and 7 : 50
(C) 7 : 22 and 7 : 53 (D) 7 : 23 and 7 : 52
(E) 7 : 21 and 7 : 49

1955 Examination

Part 1

1. Which one of the following is not equivalent to 0.000000375?

(A) 3.75×10^{-7} (B) $3\frac{3}{4} \times 10^{-7}$ (C) 375×10^{-9}

(D) $\frac{3}{8} \times 10^{-7}$ (E) $\frac{3}{8} \times 10^{-6}$

2. The smaller angle between the hands of a clock at 12:25 p.m. is:

(A) $132°30'$ (B) $137°30'$ (C) $150°$

(D) $137°32'$ (E) $137°$

3. If each number in a set of ten numbers is increased by 20, the arithmetic mean (average) of the original ten numbers:

(A) remains the same

(B) is increased by 20 (C) is increased by 200

(D) is increased by 10 (E) is increased by 2

4. The equality $\frac{1}{x-1} = \frac{2}{x-2}$ is satisfied by

(A) no real values of x (B) either $x = 1$ or $x = 2$

(C) only $x = 1$ (D) only $x = 2$

(E) only $x = 0$

5. y varies inversely as the square of x. When $y = 16$, $x = 1$. When $x = 8$, y equals:

(A) 2 (B) 128 (C) 64

(D) $\frac{1}{4}$ (E) 1024

6. A merchant buys a number of oranges at 3 for 10¢ and an equal number at 5 for 20¢. To "break even" he must sell all at:

(A) 8 for 30¢ (B) 3 for 11¢ (C) 5 for 18¢

(D) 11 for 40¢ (E) 13 for 50¢

7. If a worker receives a 20 per cent cut in wages, he may regain his original pay exactly by obtaining a raise of:

(A) 20 per cent (B) 25 per cent (C) $22\frac{1}{2}$ per cent

(D) \$20 (E) \$25

8. The graph of $x^2 - 4y^2 = 0$:

(A) is a hyperbola intersecting only the x-axis

(B) is a hyperbola intersecting only the y-axis.

(C) is a hyperbola intersecting neither axis

(D) is a pair of straight lines

(E) does not exist

9. A circle is inscribed in a triangle with sides 8, 15, and 17. The radius of the circle is:

(A) 6 (B) 2 (C) 5

(D) 3 (E) 7

10. How many hours does it take a train traveling at an average rate of 40 mph between stops to travel a miles if it makes n stops of m minutes each?

(A) $\frac{3a + 2mn}{120}$ (B) $3a + 2mn$ (C) $\frac{3a + 2mn}{12}$

(D) $\frac{a + mn}{40}$ (E) $\frac{a + 40mn}{40}$

11. The negation of the statement "No slow learners attend this school," is:

(A) All slow learners attend this school.

(B) All slow learners do not attend this school.

(C) Some slow learners attend this school.

(D) Some slow learners do not attend this school.

(E) No slow learners attend this school.

12. The solution of $\sqrt{5x-1} + \sqrt{x-1} = 2$ is:

(A) $x = 2$, $x = 1$ (B) $x = \frac{2}{3}$ (C) $x = 2$

(D) $x = 1$ (E) $x = 0$

13. The fraction $\frac{a^{-4} - b^{-4}}{a^{-2} - b^{-2}}$ is equal to:

(A) $a^{-6} - b^{-6}$ (B) $a^{-2} - b^{-2}$ (C) $a^{-2} + b^{-2}$

(D) $a^2 + b^2$ (E) $a^2 - b^2$

14. The length of rectangle R is 10 per cent more than the side of square S. The width of the rectangle is 10 per cent less than the side of the square. The ratio of the areas, $R:S$, is:

(A) 99 : 100 (B) 101 : 100 (C) 1 : 1
(D) 199 : 200 (E) 201 : 200

15. The ratio of the areas of two concentric circles is 1 : 3. If the radius of the smaller is r then the difference between the radii is best approximated by:

(A) $0.41r$ (B) 0.73 (C) 0.75
(D) $0.73r$ (E) $0.75r$

Part 2

16. The value of $\frac{3}{a+b}$ when $a = 4$ and $b = -4$ is:

(A) 3 (B) $\frac{3}{8}$ (C) 0
(D) any finite number (E) meaningless

17. If $\log x - 5 \log 3 = -2$, then x equals:

(A) 1.25 (B) 0.81 (C) 2.43
(D) 0.8 (E) either 0.8 or 1.25

18. The discriminant of the equation $x^2 + 2x\sqrt{3} + 3 = 0$ is zero. Hence, its roots are:

(A) real and equal (B) rational and equal
(C) rational and unequal (D) irrational and unequal
(E) imaginary

19. Two numbers whose sum is 6 and the absolute value of whose difference is 8 are roots of the equation:

(A) $x^2 - 6x + 7 = 0$ (B) $x^2 - 6x - 7 = 0$
(C) $x^2 + 6x - 8 = 0$ (D) $x^2 - 6x + 8 = 0$
(E) $x^2 + 6x - 7 = 0$

20. The expression $\sqrt{25 - t^2} + 5$ equals zero for:

(A) no real or imaginary values of t
(B) no real values of t, but for some imaginary values
(C) no imaginary values of t, but for some real values
(D) $t = 0$
(E) $t = \pm 5$

21. Represent the hypotenuse of a right triangle by c and the area by A. The altitude on the hypotenuse is:

(A) $\frac{A}{c}$ (B) $\frac{2A}{c}$ (C) $\frac{A}{2c}$

(D) $\frac{A^2}{c}$ (E) $\frac{A}{c^2}$

22. On a \$10,000 order a merchant has a choice between three successive discounts of 20%, 20%, and 10% and three successive discounts of 40%, 5%, and 5%. By choosing the better offer, he can save:

(A) nothing at all (B) \$400 (C) \$330

(D) \$345 (E) \$360

23. In checking the petty cash a clerk counts q quarters, d dimes, n nickels, and c cents. Later he discovers that x of the nickels were counted as quarters and x of the dimes were counted as cents. To correct the total obtained the clerk must:

(A) make no correction (B) subtract 11¢ (C) subtract $11x$¢

(D) add $11x$¢ (E) add x¢

24. The function $4x^2 - 12x - 1$:

(A) always increases as x increases

(B) always decreases as x decreases to 1

(C) cannot equal 0

(D) has a maximum value when x is negative

(E) has a minimum value of -10

25. One of the factors of $x^4 + 2x^2 + 9$ is:

(A) $x^2 + 3$ (B) $x + 1$ (C) $x^2 - 3$

(D) $x^2 - 2x - 3$ (E) none of these

26. Mr. A owns a house worth \$10,000. He sells it to Mr. B at 10% profit. Mr. B sells the house back to Mr. A at a 10% loss. Then:

(A) A comes out even (B) A makes \$100 (C) A makes \$1,000

(D) B loses \$100 (E) none of these is correct

27. If r and s are the roots of $x^2 - px + q = 0$, then $r^2 + s^2$ equals:

(A) $p^2 + 2q$ (B) $p^2 - 2q$ (C) $p^2 + q^2$

(D) $p^2 - q^2$ (E) p^2

28. On the same set of axes are drawn the graph of $y = ax^2 + bx + c$ and the graph of the equation obtained by replacing x by $-x$ in the given equation. If $b \neq 0$ and $c \neq 0$ these two graphs intersect:

(A) in two points, one on the x-axis and one on the y-axis

(B) in one point located on neither axis

(C) only at the origin

(D) in one point on the x-axis

(E) in one point on the y-axis

29. In the figure PA is tangent to semicircle SAR; PB is tangent to semicircle RBT; SRT is a straight line; the arcs are indicated in the figure. Angle APB is measured by:

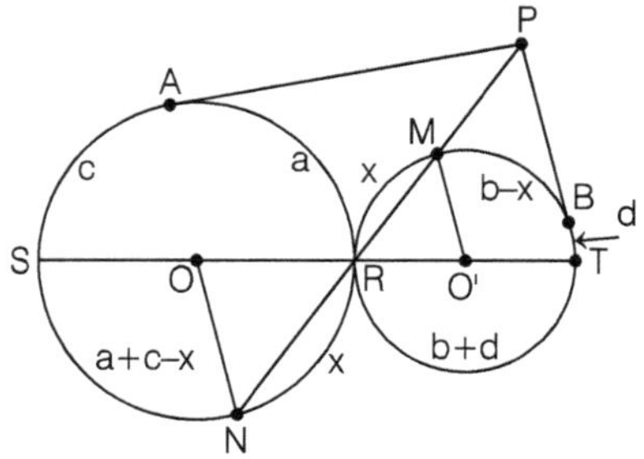

(A) $\frac{1}{2}(a-b)$ (B) $\frac{1}{2}(a+b)$ (C) $(c-a)-(d-b)$

(D) $a-b$ (E) $a+b$

30. Each of the equations $3x^2-2=25$, $(2x-1)^2=(x-1)^2$, $\sqrt{x^2-7}=\sqrt{x-1}$ has:

(A) two integral roots

(B) no root greater than 3

(C) no root zero

(D) only one root

(E) one negative root and one positive root

31. An equilateral triangle whose side is 2 is divided into a triangle and a trapezoid by a line drawn parallel to one of its sides. If the area of the trapezoid equals one-half of the area of the original triangle, the length of the median of the trapezoid is:

(A) $\frac{\sqrt{6}}{2}$ (B) $\sqrt{2}$ (C) $2+\sqrt{2}$

(D) $\frac{2+\sqrt{2}}{2}$ (E) $\frac{2\sqrt{3}-\sqrt{6}}{2}$

32. If the discriminant of $ax^2+2bx+c=0$ is zero, then another true statement about a, b and c is that:

(A) they form an arithmetic progression

(B) they form a geometric progression

(C) they are unequal

(D) they are all negative numbers

(E) only b is negative and a and c are positive

33. Henry starts a trip when the hands of the clock are together between 8 a.m. and 9 a.m. He arrives at his destination between 2 p.m. and 3 p.m. when the hands of the clock are exactly 180° apart. The trip takes:

(A) 6 hr. (B) 6 hr $43\frac{7}{11}$ min. (C) 5 hr $16\frac{4}{11}$ min.

(D) 6 hr 30 min. (E) none of these

34. A 6-inch and 18-inch diameter pole are placed as in the figure and bound together with wire. The length of the shortest wire that will go around them is:

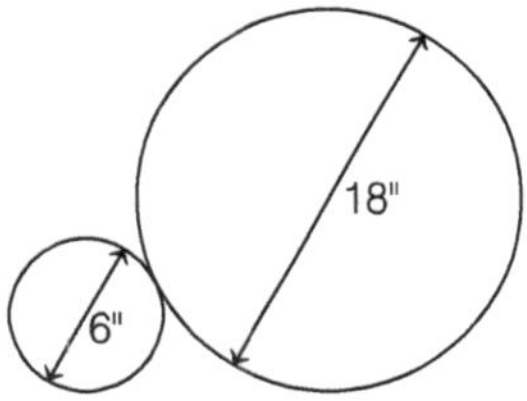

(A) $12\sqrt{3}+16\pi$ (B) $12\sqrt{3}+7\pi$ (C) $12\sqrt{3}+14\pi$

(D) $12+15\pi$ (E) 24π

35. Three boys agree to divide a bag of marbles in the following manner. The first boy takes one more than half the marbles. The second takes a third of the number remaining. The third boy finds that he is left with twice as many marbles as the second boy. The original number of marbles:

(A) is 8 or 38 (B) cannot be determined from the given data

(C) is 20 or 26 (D) is 14 or 32

(E) is none of these

Part 3

36. A cylindrical oil tank, lying horizontally, has an interior length of 10 feet and an interior diameter of 6 feet. If the rectangular surface of the oil has an area of 40 square feet, the depth of the oil is:

(A) $\sqrt{5}$ (B) $2\sqrt{5}$ (C) $3-\sqrt{5}$

(D) $3+\sqrt{5}$ (E) either $3-\sqrt{5}$ or $3+\sqrt{5}$

37. A three-digit number has, from left to right, the digits h, t, and u with $h > u$. When the number with the digits reversed is subtracted from the original number, the units' digit in the difference is 4. The next two digits, from right to left, are:

(A) 5 and 9 (B) 9 and 5 (C) impossible to tell

(D) 5 and 4 (E) 4 and 5

38. Four positive integers are given. Select any three of these integers, find their arithmetic average, and add this result to the fourth integer. Thus the numbers 29, 23, 21 and 17 are obtained. One of the original integers is:

(A) 19 (B) 21 (C) 23
(D) 29 (E) 17

39. If $y = x^2 + px + q$, then if the least possible value of y is zero, q is equal to:

(A) 0 (B) $\frac{p^2}{4}$ (C) $\frac{p}{2}$
(D) $-\frac{p}{2}$ (E) $\frac{p^2}{4} - q$

40. If $b \neq d$, the fractions $\frac{ax+b}{cx+d}$ and $\frac{b}{d}$ are unequal if:

(A) $a = c = 1$ and $x \neq 0$ (B) $a = b = 0$ (C) $a = c = 0$
(D) $x = 0$ (E) $ad = bc$

41. A train traveling from Aytown to Beetown meets with an accident after 1 hr. It is stopped for $\frac{1}{2}$ hr. after which it proceeds at four-fifths of its usual rate, arriving at Beetown 2 hr. late. If the train had covered 80 miles more before the accident, it would have been just 1 hr. late. The usual rate of the train is:

(A) 20 mph (B) 30 mph (C) 40 mph
(D) 50 mph (E) 60 mph

42. If a, b, and c are positive integers, the radicals $\sqrt{a + \frac{b}{c}}$ and $a\sqrt{\frac{b}{c}}$ are equal when and only when:

(A) $a = b = c = 1$ (B) $a = b$ and $c = a = 1$
(C) $c = \frac{b(a^2-1)}{a}$ (D) $a = b$ and c is any value
(E) $a = b$ and $c = a - 1$

43. The pairs of values of x and y that are the common solutions of the equations $y = (x+1)^2$ and $xy + y = 1$ are:

(A) 3 real pairs
(B) 4 real pairs
(C) 4 imaginary pairs
(D) 2 real and 2 imaginary pairs
(E) 1 real and 2 imaginary pairs

44. In circle O chord AB is produced so that $\overline{BC}$ equals a radius of the circle. CO is drawn and extended to D. AO is drawn. Which of the following expresses the relationship between angles x and y?

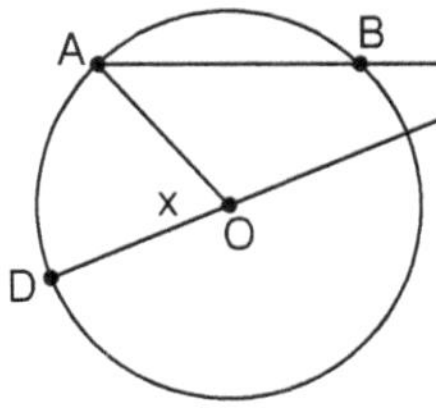

(A) $x = 3y$ (B) $x = 2y$ (C) $x = 60°$
(D) there is no special relationship between x and y
(E) $x = 2y$ or $x = 3y$, depending upon the length of AB

45. Given a geometric sequence with the first term $\neq 0$ and $r \neq 0$ and an arithmetic sequence with the first term $= 0$. A third sequence 1, 1, 2, ... is formed by adding corresponding terms of the two given sequences. The sum of the first ten terms of the third sequence is:
(A) 978 (B) 557 (C) 467
(D) 1068
(E) not possible to determine from the information given

46. The graphs of $2x + 3y - 6 = 0$, $4x - 3y - 6 = 0$, $x = 2$, and $y = \frac{2}{3}$ intersect in:
(A) 6 points (B) 1 point (C) 2 points
(D) no points (E) an unlimited number of points

47. The expressions $a + bc$ and $(a + b)(a + c)$ are:
(A) always equal (B) never equal
(C) equal when $a + b + c = 1$ (D) equal when $a + b + c = 0$
(E) equal only when $a = b = c = 0$

48. Given triangle ABC with medians AE, BF, CD, FH parallel and equal in length to AE, BH and HE are drawn, FE extended meets BH in G. Which one of the following statements is not necessarily correct?

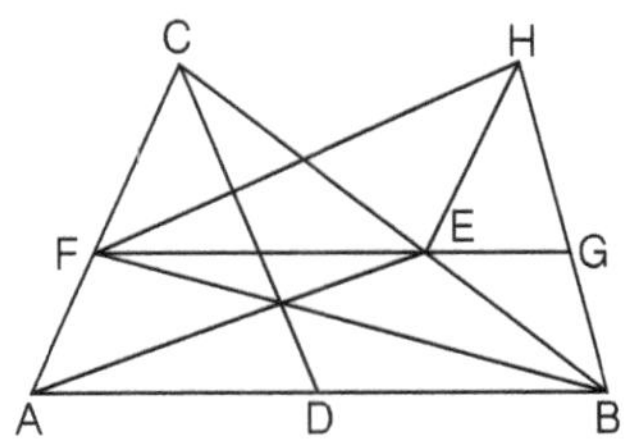

(A) $AEHF$ is parallelogram
(B) $\overline{HE} = \overline{HG}$
(C) $\overline{BH} = \overline{DC}$
(D) $\overline{FG} = \frac{3}{4}\overline{AB}$
(E) FG is a median of triangle BFH

49. The graphs of $y = \dfrac{x^2 - 4}{x - 2}$ and $y = 2x$ intersect in:

(A) one point whose abscissa is 2
(B) one point whose abscissa is 0
(C) no points
(D) two distinct points
(E) two identical points

50. In order to pass B going 40 mph on a two-lane highway A, going 50 mph, must gain 30 feet. Meantime, C, 210 feet from A, is headed toward him at 50 mph. If B and C maintain their speeds, then, in order to pass safely, A must increase his speed by:

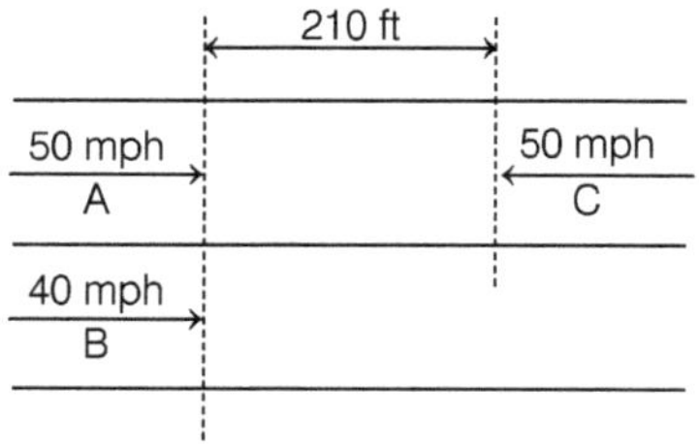

This figure is not drawn to scale

(A) 30 mph
(B) 10 mph
(C) 5 mph
(D) 15 mph
(E) 3 mph

1956 Examination

Part 1

1. The value of $x + x(x^x)$ when $x = 2$ is:

   (A) 10 (B) 16 (C) 18
   (D) 36 (E) 64

2. Mr. Jones sold two pipes at \$1.20 each. Based on the cost his profit on one was 20% and his loss on the other was 20%. On the sale of the pipes, he:

   (A) broke even (B) lost 4¢ (C) gained 4¢
   (D) lost 10¢ (E) gained 10¢

3. The distance light travels in one year is approximately 5,870,000,000,000 miles. The distance light travels in 100 years is:

   (A) 587×10^8 miles (B) 587×10^{10} miles (C) 587×10^{-10} miles
   (D) 587×10^{12} miles (E) 587×10^{-12} miles

4. A man has \$10,000 to invest. He invests \$4,000 at 5% and \$3,500 at 4%. In order to have a yearly income of \$500, he must invest the remainder at:

   (A) 6% (B) 6.1% (C) 6.2%
   (D) 6.3% (E) 6.4%

5. A nickel is placed on a table. The number of nickels which can be placed around it, each tangent to it and to two others is:

   (A) 4 (B) 5 (C) 6
   (D) 8 (E) 12

6. In a group of cows and chickens, the number of legs was 14 more than twice the number of heads. The number of cows was:

   (A) 5 (B) 7 (C) 10
   (D) 12 (E) 14

7. The roots of the equation $ax^2 + bx + c = 0$ will be reciprocal if:

   (A) $a = b$ (B) $a = bc$ (C) $c = a$
   (D) $c = b$ (E) $c = ab$

8. If $8 \cdot 2^x = 5^{y+8}$, then , when $y = -8$, $x =$

(A) -4 (B) -3 (C) 0
(D) 4 (E) 8

9. Simplify $[\sqrt[3]{\sqrt[6]{a^9}}]^4\ [\sqrt[6]{\sqrt[3]{a^9}}]^4$, the result is:

(A) a^{16} (B) a^{12} (C) a^8
(D) a^4 (E) a^2

10. A circle of radius 10 inches has its center at the vertex C of an equilateral triangle ABC and passes through the other two vertices. The side AC extended through C intersects the circle at D. The number of degrees of angle ADB is:

(A) 15 (B) 30 (C) 60
(D) 90 (E) 120

11. The expression $1 - \frac{1}{1+\sqrt{3}} + \frac{1}{1-\sqrt{3}}$ equals:

(A) $1-\sqrt{3}$ (B) 1 (C) $-\sqrt{3}$
(D) $\sqrt{3}$ (E) $1+\sqrt{3}$

12. If $x^{-1} - 1$ is divided by $x - 1$ the quotient is:

(A) 1 (B) $\frac{1}{x-1}$ (C) $\frac{-1}{x-1}$
(D) $\frac{1}{x}$ (E) $-\frac{1}{x}$

13. Given two positive integers x and y with $x < y$. The percent that x is less than y is:

(A) $\frac{100(y-x)}{x}$ (B) $\frac{100(x-y)}{x}$
(C) $\frac{100(y-x)}{y}$ (D) $100(y-x)$
(E) $100(x-y)$

14. The points $A, B,$ and C are on a circle O. The tangent line at A and the secant BC intersect at P, B lying between C and P. If $\overline{BC} = 20$ and $\overline{PA} = 10\sqrt{3}$, then $\overline{PB}$ equals:

(A) 5 (B) 10 (C) $10\sqrt{3}$
(D) 20 (E) 30

15. The root(s) of $\frac{15}{x^2-4} - \frac{2}{x-2} = 1$ is (are):

(A) -5 and 3 (B) ± 2 (C) 2 only
(D) -3 and 5 (E) 3 only

Part 2

16. The sum of three numbers is 98. The ratio of the first to the second is $\frac{2}{3}$, and the ratio of the second to the third is $\frac{5}{8}$. The second number is:

(A) 15 (B) 20 (C) 30
(D) 32 (E) 33

17. The fraction $\frac{5x-11}{2x^2+x-6}$ was obtained by adding the two fractions $\frac{A}{x+2}$ and $\frac{B}{2x-3}$. The values of A and B must be:

(A) $A = 5x$, $B = -11$, (B) $A = -11$, $B = 5x$ (C) $A = -1$ $B = 3$
(D) $A = 3$ $B = -1$ (E) $A = 5$ $B = -11$

18. If $10^{2v} = 25$, then 10^{-v} equals:

(A) $-\frac{1}{5}$ (B) $\frac{1}{625}$ (C) $\frac{1}{50}$
(D) $\frac{1}{25}$ (E) $\frac{1}{5}$

19. Two candles of the same height are lighted at the same time. The first is consumed in 4 hours and the second in 3 hours. Assuming that each candle burns at a constant rate, in how many hours after being lighted was the first candle twice the height of the second?

(A) $\frac{3}{4}$ hr. (B) $1\frac{1}{2}$ hr. (C) 2 hr.
(D) $2\frac{2}{5}$ hr. (E) $2\frac{1}{2}$ hr.

20. If $(0.2)^x = 2$ and $\log 2 = 0.3010$, then the value of x to the nearest tenth is:

(A) – 10.0 (B) – 0.5 (C) – 0.4
(D) – 0.2 (E) 10.0

21. If each of two intersecting lines intersects a hyperbola and neither line is tangent to the hyperbola, then the possible number of points of intersection with the hyperbola is:

(A) 2 (B) 2 or 3
(C) 2 or 4 (D) 3 or 4
(E) 2, 3, or 4

22. Jones covered a distance of 50 miles on his first trip. On a later trip he traveled 300 miles while going three times as fast. His new time compared with the old time was:

(A) three times as much (B) twice as much
(C) the same (D) half as much
(E) a third as much

23. About the equation $ax^2 - 2x\sqrt{2} + c = 0$, with a and c real constants, we are told that the discriminant is zero. The roots are necessarily:

(A) equal and integral (B) equal and rational
(C) equal and real (D) equal and irrational
(E) equal and imaginary

24. In the figure $\overline{AB} = \overline{AC}$, angle $BAD = 30°$, and $\overline{AE} = \overline{AD}$. Then x equals:

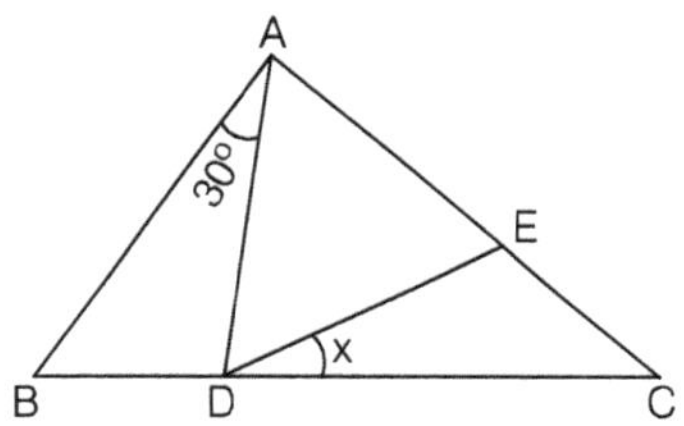

(A) $7\frac{1}{2}^\circ$ (B) 10° (C) $12\frac{1}{2}^\circ$
(D) 15° (E) 20°

25. The sum of all numbers of the form $2k + 1$, where k takes on integral values from 1 to n is:

(A) n^2 (B) $n(n + 1)$ (C) $n(n + 2)$
(D) $(n + 1)^2$ (E) $(n + 1)(n + 2)$

26. Which one of the following combinations of given parts does not determine the indicated triangle?

(A) base angle and vertex angle; isosceles triangle
(B) vertex angle and the base; isosceles triangle
(C) the radius of the circumscribed circle; equilateral triangle
(D) one arm and the radius of the inscribed circle; right triangle
(E) two angles and a side opposite one of them; scalene triangle

27. If an angle of a triangle remains unchanged but each of its two including sides is doubled, then the area is multiplied by:

(A) 2 (B) 3 (C) 4
(D) 6 (E) more than 6

28. Mr. J left his entire estate to his wife, his daughter, his son, and the cook. His daughter and son got half the estate, sharing in the ratio of 4 to 3. His wife got twice as much as the son. If the cook received a bequest of \$500, then the entire estate was:

(A) \$3500 (B) \$5500 (C) \$6500
(D) \$7000 (E) \$7500

29. The points of intersection of $xy = 12$ and $x^2 + y^2 = 25$ are joined in succession. The resulting figure is:

(A) a straight line (B) an equilateral triangle
(C) a parallelogram (D) a rectangle
(E) a square

30. If the altitude of an equilateral triangle is $\sqrt{6}$, then the area is:

(A) $2\sqrt{2}$ (B) $2\sqrt{3}$ (C) $3\sqrt{3}$
(D) $6\sqrt{2}$ (E) 12

31. In our number system the base is ten. If the base were changed to four you would count as follows : 1, 2, 3, 10, 11, 12, 13, 20, 21, 22, 23, 30, ... The twentieth number would be:

(A) 20 (B) 38 (C) 44
(D) 104 (E) 110

32. George and Henry started a race from opposite ends of the pool. After a minute and a half, they passed each other in the center of the pool. If they lost no time in turning and maintained their respective speeds, how many minutes after starting did they pass each other the second time?

(A) 3 (B) $4\frac{1}{2}$ (C) 6
(D) $7\frac{1}{2}$ (E) 9

33. The number $\sqrt{2}$ is equal to:

(A) a rational fraction (B) a finite decimal (C) 1.41421
(D) an infinite repeating decimal
(E) an infinite non-repeating decimal

34. If n is any whole number, $n^2(n^2 - 1)$ is always divisible by:

(A) 12 (B) 24 (C) any multiple of 12
(D) $12 - n$ (E) 12 and 24

35. A rhombus is formed by two radii and two chords of a circle whose radius is 16 feet. The area of the rhombus in square feet is:

(A) 128 (B) $128\sqrt{3}$ (C) 256
(D) 512 (E) $512\sqrt{3}$

Part 3

36. If the sum $1 + 2 + 3 + \dots + K$ is a perfect square N^2 and if N is less than 100, then the possible values for K are:

(A) only 1 (B) 1 and 8 (C) only 8
(D) 8 and 49 (E) 1, 8, and 49

37. On a map whose scale is 400 miles to an inch and a half, a certain estate is represented by a rhombus having a 60° angle. The diagonal opposite 60° is $\frac{3}{16}$ in. The area of the estate in square miles is:

(A) $\frac{2500}{\sqrt{3}}$ (B) $\frac{1250}{\sqrt{3}}$ (C) 1250
(D) $\frac{5625\sqrt{3}}{2}$ (E) $1250\sqrt{3}$

38. In a right triangle with sides a and b, and hypotenuse c, the altitude drawn on the hypotenuse is x. Then:

(A) $a \cdot b = x^2$ (B) $\frac{1}{a} + \frac{1}{b} = \frac{1}{x}$ (C) $a^2 + b^2 = 2x^2$
(D) $\frac{1}{x^2} = \frac{1}{a^2} + \frac{1}{b^2}$ (E) $\frac{1}{x} = \frac{b}{a}$

39. The hypotenuse c and one arm a of a right triangle are consecutive integers. The square of the second arm is:

(A) ca (B) $\frac{c}{a}$ (C) $c + a$
(D) $c - a$ (E) none of these

40. If $V = gt + V_0$ and $S = \frac{1}{2}gt^2 + V_0t$, then t equals:

(A) $\frac{2S}{V + V_0}$ (B) $\frac{2S}{V - V_0}$ (C) $\frac{2S}{V_0 - V}$
(D) $\frac{2S}{V}$ (E) $2S - V$

41. The equation $3y^2 + y + 4 = 2(6x^2 + y + 2)$ where $y = 2x$ is satisfied by:

(A) no value of x (B) all values of x (C) $x = 0$ only
(D) all integral values of x only
(E) all rational values of x only

42. The equation $\sqrt{x + 4} - \sqrt{x - 3} + 1 = 0$ has:

(A) no root (B) one real root
(C) one real root and one imaginary root
(D) two imaginary roots (E) two real roots

43. The number of scalene triangles having all sides of integral lengths, and perimeter less than 13 is:

(A) 1 (B) 2 (C) 3
(D) 4 (E) 18

44. If $x < a < 0$ means that x and a are numbers such that x is less than a and a is less than zero, then:

(A) $x^2 < ax < 0$ (B) $x^2 > ax > a^2$ (C) $x^2 < a^2 < 0$
(D) $x^2 > ax$ but $ax < 0$ (E) $x^2 > a^2$ but $a^2 < 0$

45. A wheel with a rubber tire has an outside diameter of 25 in. When the radius has been decreased a quarter of an inch, the number of revolutions of the wheel in one mile will:

(A) be increased about 2%
(B) be increased about 1%
(C) be increased about 20%
(D) be increased $\frac{1}{2}$%
(E) remain the same

46. For the equation $\frac{1+x}{1-x} = \frac{N+1}{N}$ to be true where N is positive, x can have:

(A) any positive value less than 1
(B) any value less than 1
(C) the value zero only
(D) any non-negative value
(E) any value

47. An engineer said he could finish a highway section in 3 days with his present supply of a certain type of machine. However, with 3 more of these machines the job could be done in 2 days. If the machines all work at the same rate, how many days would it take to do the job with one machine?

(A) 6 (B) 12 (C) 15
(D) 18 (E) 36

48. If p is a positive integer, then $\frac{3p+25}{2p-5}$ can be a positive integer, if and only if p is:

(A) at least 3 (B) at least 3 and no more than 35
(C) no more than 35 (D) equal to 35
(E) equal to 3 or 35

49. Triangle PAB is formed by three tangents to circle O and $\measuredangle APB = 40°$; then angle AOB equals:

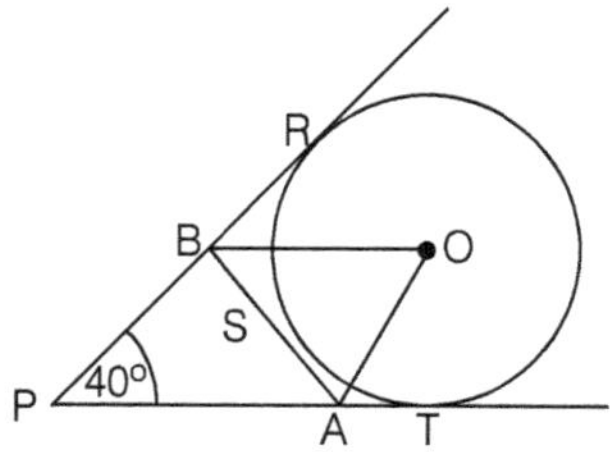

(A) 45° (B) 50° (C) 55°
(D) 60° (E) 70°

50. In triangle ABC, $\overline{CA} = \overline{CB}$. On CB square $BCDE$ is constructed away from the triangle. If x is the number of degrees in angle DAB, then

(A) x depends upon triangle ABC
(B) x is independent of the triangle
(C) x may equal angle CAD
(D) x can never equal angle CAB
(E) x is greater than 45° but less than 90°

1957 Examination

Part 1

1. The number of distinct lines representing the altitudes, medians, and interior angle bisectors of a triangle that is isosceles, but not equilateral, is:

(A) 9 (B) 7 (C) 6
(D) 5 (E) 3

2. In the equation $2x^2 - hx + 2k = 0$, the sum of the roots is 4 and the product of the roots is -3. Then h and k have the values, respectively:

(A) 8 and -6 (B) 4 and -3 (C) -3 and 4
(D) -3 and 8 (E) 8 and -3

3. The simplest form of $1 - \dfrac{1}{1 + \dfrac{a}{1-a}}$ is:

(A) a if $a \neq 0$ (B) 1 (C) a if $a \neq -1$
(D) $1 - a$ with no restriction on a
(E) a if $a \neq 1$

4. The first step in finding the product $(3x + 2)(x - 5)$ by use of the distributive property in the form $a(b + c) = ab + ac$ is:

(A) $3x^2 - 13x - 10$ (B) $3x(x - 5) + 2(x - 5)$
(C) $(3x + 2)x + (3x + 2)(-5)$ (D) $3x^2 - 17x - 10$
(E) $3x^2 + 2x - 15x - 10$

5. Through the use of theorems on logarithms

$$\log\frac{a}{b} + \log\frac{b}{c} + \log\frac{c}{d} - \log\frac{ay}{dx}$$

can be reduced to:

(A) $\log\dfrac{y}{x}$ (B) $\log\dfrac{x}{y}$ (C) 1
(D) 0 (E) $\log\dfrac{a^2y}{d^2x}$

6. An open box is constructed by starting with a rectangular sheet of metal 10 in. by 14 in. and cutting a square of side x inches from each corner. The resulting projections are folded up and the seams welded. The volume of the resulting box is:

(A) $140x - 48x^2 + 4x^3$ (B) $140x + 48x^2 + 4x^3$
(C) $140x + 24x^2 + x^3$ (D) $140x - 24x^2 + x^3$
(E) none of these

7. The area of a circle inscribed in an equilateral triangle is 48π. The perimeter of this triangle is:

(A) $72\sqrt{3}$ (B) $48\sqrt{3}$ (C) 36
(D) 24 (E) 72

8. The numbers x, y, z are proportional to 2, 3, 5. The sum of x, y and z is 100. The number y is given by the equation $y = ax - 10$. Then a is:

(A) 2 (B) $\frac{3}{2}$ (C) 3
(D) $\frac{5}{2}$ (E) 4

9. The value of $x - y^{x-y}$ when $x = 2$ and $y = -2$ is:

(A) -18 (B) -14 (C) 14
(D) 18 (E) 256

10. The graph of $y = 2x^2 + 4x + 3$ has its:

(A) lowest point at $(-1, 9)$
(B) lowest point at $(1, 1)$
(C) lowest point at $(-1, 1)$
(D) highest point at $(-1, 9)$
(E) highest point at $(-1, 1)$

11. The angle formed by the hands of a clock at 2:15 is:

(A) $30°$ (B) $27\frac{1}{2}°$ (C) $157\frac{1}{2}°$
(D) $172\frac{1}{2}°$ (E) none of these

12. Comparing the numbers 10^{-49} and $2 \cdot 10^{-50}$ we may say:

(A) the first exceeds the second by $8 \cdot 10^{-1}$
(B) the first exceeds the second by $2 \cdot 10^{-1}$
(C) the first exceeds the second by $8 \cdot 10^{-50}$
(D) the second is five times the first
(E) the first exceeds the second by 5

13. A rational number between $\sqrt{2}$ and $\sqrt{3}$ is:

(A) $\frac{\sqrt{2}+\sqrt{3}}{2}$ (B) $\frac{\sqrt{2}\cdot\sqrt{3}}{2}$ (C) 1.5

(D) 1.8 (E) 1.4

14. If $y=\sqrt{x^2-2x+1}+\sqrt{x^2+2x+1}$, then y is:

(A) $2x$ (B) $2(x+1)$ (C) 0

(D) $|x-1|+|x+1|$ (E) none of these

15. The table below shows the distance s in feet a ball rolls down an inclined plane in t seconds.

t	0	1	2	3	4	5
s	0	10	40	90	160	250

The distance s for $t=2.5$ is:

(A) 45 (B) 62.5 (C) 70

(D) 75 (E) 82.5

Part 2

16. Goldfish are sold at 15¢ each. The rectangular coordinate graph showing the cost of 1 to 12 goldfish is:

(A) a straight line segment

(B) a set of horizontal parallel line segments

(C) a set of vertical parallel line segments

(D) a finite set of distinct points

(E) a straight line

17. A cube is made by soldering twelve 3-inch lengths of wire properly at the vertices of the cube. If a fly alights at one of the vertices and then walks along the edges, the greatest distance it could travel before coming to any vertex a second time, without retracing any distance, is:

(A) 24 in. (B) 12 in. (C) 30 in.

(D) 18 in. (E) 36 in.

18. Circle O has diameters AB and CD perpendicular to each other. AM is any chord intersecting CD at P. Then $\overline{AP}\cdot\overline{AM}$ is equal to:

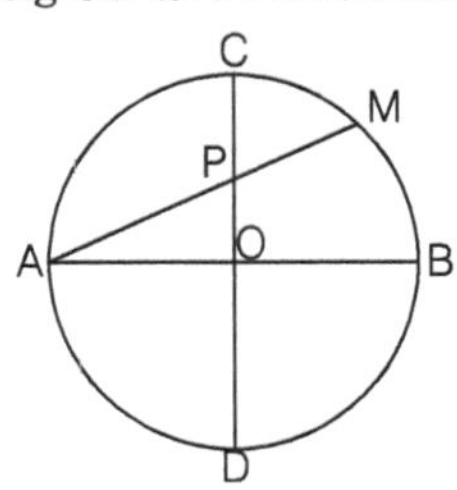

(A) $\overline{AO}\cdot\overline{OB}$ (B) $\overline{AO}\cdot\overline{AB}$ (C) $\overline{CP}\cdot\overline{CD}$
(D) $\overline{CP}\cdot\overline{PD}$ (E) $\overline{CO}\cdot\overline{OP}$

19. The base of the decimal number system is ten, meaning, for example, that $123 = 1\cdot 10^2 + 2\cdot 10 + 3$. In the binary system, which has base two, the first five positive integers are 1, 10, 11, 100, 101. The numeral 10011 in the binary system would then be written in the decimal system as:

(A) 19 (B) 40 (C) 10011
(D) 11 (E) 7

20. A man makes a trip by automobile at an average speed of 50 mph. He returns over the same route at an average speed of 45 mph. His average speed for the entire trip is:

(A) $47\frac{7}{19}$ (B) $47\frac{1}{4}$ (C) $47\frac{1}{2}$
(D) $47\frac{11}{19}$ (E) none of these

21. Start with the theorem "If two angles of a triangle are equal, the triangle is isosceles," and the following four statements:

1. If two angles of a triangle are not equal, the triangle is not isosceles.
2. The base angles of an isosceles triangle are equal.
3. If a triangle is not isosceles, then two of its angles are not equal.
4. A necessary condition that two angles of a triangle be equal is that the triangle be isosceles.

Which combination of statements contains only those which are logically equivalent to the given theorem?

(A) 1, 2, 3, 4 (B) 1, 2, 3 (C) 2, 3, 4
(D) 1, 2 (E) 3, 4

22. If $\sqrt{x-1} - \sqrt{x+1} + 1 = 0$, then $4x$ equals:

(A) 5 (B) $4\sqrt{-1}$ (C) 0
(D) $1\frac{1}{4}$ (E) no real value

23. The graph of $x^2 + y = 10$ and the graph of $x + y = 10$ meet in two points. The distance between these two points is:

(A) less than 1 (B) 1 (C) $\sqrt{2}$
(D) 2 (E) more than 2

24. If the square of a number of two digits is decreased by the square of the number formed by reversing the digits, then the result is not always divisible by:

(A) 9 (B) the product of the digits (C) the sum of the digits
(D) the difference of the digits (E) 11

25. The vertices of triangle PQR have coordinates as follows : $P(0, a)$, $Q(b, 0)$, $R(c, d)$, where a, b, c and d are positive. The origin and point R lie on opposite sides of PQ. The area of triangle PQR may be found from the expression:

(A) $\frac{ab + ac + bc + cd}{2}$ (B) $\frac{ac + bd - ab}{2}$

(C) $\frac{ab - ac - bd}{2}$ (D) $\frac{ac + bd + ab}{2}$

(E) $\frac{ac + bd - ab - cd}{2}$

26. From a point within a triangle, line segments are drawn to the vertices. A necessary and sufficient condition that the three triangles thus formed have equal areas is that the point be:

(A) the center of the inscribed circle

(B) the center of the circumscribed circle

(C) such that the three angles formed at the point each be 120°

(D) the intersection of the altitudes of the triangle

(E) the intersection of the medians of the triangle

27. The sum of the reciprocals of the roots of the equation $x^2 + px + q = 0$ is:

(A) $-p/q$ (B) q/p (C) p/q

(D) $-q/p$ (E) pq

28. If a and b are positive and $a \neq 1$, $b \neq 1$, then the value of $b^{\log_b a}$ is:

(A) dependent upon b

(B) dependent upon a

(C) dependent upon a and b

(D) zero

(E) one

29. The relation $x^2(x^2 - 1) \geq 0$ is true only for:

(A) $x \geq 1$ (B) $-1 \leq x \leq 1$

(C) $x = 0$, $x = 1$, $x = -1$ (D) $x = 0$, $x \leq -1$, $x \geq 1$

(E) $x \geq 0$

where $x \geq a$ means that x can take on all values greater than a and the value equal to a, while $x \leq a$ has a corresponding meaning with "less than."

30. The sum of the squares of the first n positive integers is given by the expression $\frac{n(n + c)(2n + k)}{6}$, if c and k are, respectively:

(A) 1 and 2 (B) 3 and 5 (C) 2 and 2

(D) 1 and 1 (E) 2 and 1

31. A regular octagon is to be formed by cutting equal isosceles right triangles from the corners of a square. If the square has sides of one unit, the leg of each of the triangles has length:

(A) $\frac{2+\sqrt{2}}{3}$ (B) $\frac{2-\sqrt{2}}{2}$ (C) $\frac{1+\sqrt{2}}{2}$

(D) $\frac{1+\sqrt{2}}{3}$ (E) $\frac{2-\sqrt{2}}{3}$

32. The largest of the following integers which divides each of the members of the sequence $1^5-1, 2^5-2, 3^5-3, \ldots, n^5-n, \ldots$ is:

(A) 1 (B) 60 (C) 15

(D) 120 (E) 30

33. If $9^{x+2} = 240 + 9^x$, then the value of x is:

(A) 0.1 (B) 0.2 (C) 0.3

(D) 0.4 (E) 0.5

34. The points that satisfy the system $x + y = 1$, $x^2 + y^2 < 25$, where the symbol "<" means "less than," constitute the following set:

(A) only two points

(B) an arc of a circle

(C) a straight line segment not including the end-points

(D) a straight line segment including the end-points

(E) a single point

35. Side AC of right triangle ABC is divided into 8 equal parts. Seven line segments parallel to BC are drawn to AB from the points of division. If $BC = 10$, then the sum of the lengths of the seven line segments:

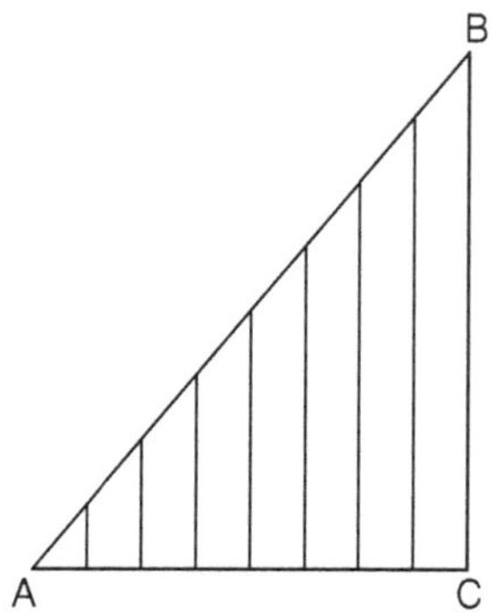

(A) can not be found from the given information

(B) is 33 (C) is 34 (D) is 35

(E) is 45

Part 3

36. If $x + y = 1$, then the largest value of xy is:

(A) 1 (B) 0.5

(C) an irrational number about 0.4

(D) 0.25 (E) 0

37. In right triangle ABC, $\overline{BC} = 5$, $\overline{AC} = 12$, and $\overline{AM} = x$; $MN \perp AC$, $NP \perp BC$; N is on AB. If $y = \overline{MN} + \overline{NP}$, one-half the perimeter of rectangle $MCPN$, then:

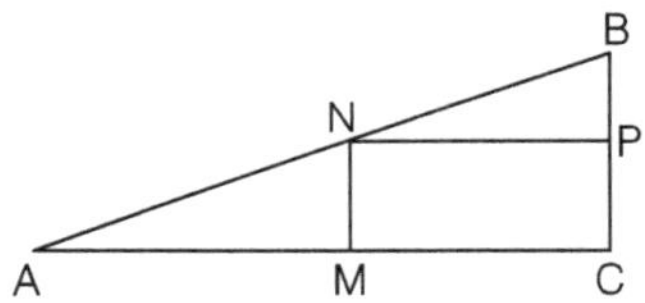

(A) $y = \frac{1}{2}(5 + 12)$ (B) $y = \frac{5x}{12} + \frac{12}{5}$

(C) $y = \frac{144 - 7x}{12}$ (D) $y = 12$

(E) $y = \frac{5x}{12} + 6$

38. From a two-digit number N we subtract the number with the digits reversed and find that the result is a positive perfect cube. Then:

(A) N cannot end in 5 (B) N can end in any digit other than 5

(C) N does not exist (D) there are exactly 7 values for N

(E) there are exactly 10 values for N

39. Two men set out at the same time to walk towards each other from M and N, 72 miles apart. The first man walks at the rate of 4 mph. The second man walks 2 miles the first hour, $2\frac{1}{2}$ miles the second hour, 3 miles the third hour, and so on in arithmetic progression. Then, the men will meet:

(A) in 7 hours

(B) in $8\frac{1}{4}$ hours

(C) nearer M than N

(D) nearer N than M

(E) midway between M and N

40. If the parabola $y = -x^2 + bx - 8$ has its vertex on the x-axis, then b must be:

(A) a positive integer

(B) a positive or a negative rational number

(C) a positive rational number

(D) a positive or a negative irrational number

(E) a negative irrational number

41. Given the system of equations $\begin{matrix} ax + (a-1)y = 1 \\ (a+1)x - ay = 1 \end{matrix}$. For which one of the following values of a is there no solution for x and y?

(A) 1 (B) 0

(C) -1 (D) $\frac{\pm\sqrt{2}}{2}$

(E) $\pm\sqrt{2}$

42. If $S = i^n + i^{-n}$, where $i = \sqrt{-1}$ and n is an integer, then the total number of possible distinct values for S is:

(A) 1 (B) 2 (C) 3

(D) 4 (E) more than 4

43. We define a lattice point as a point whose coordinates are integers, zero admitted. Then the number of lattice points on the boundary and inside the region bounded by the x-axis, the line $x = 4$, and the parabola $y = x^2$ is:

(A) 24 (B) 35 (C) 34

(D) 30 (E) not finite

44. In triangle ABC, $\overline{AC} = \overline{CD}$ and $\measuredangle CAB - \measuredangle ABC = 30°$. Then, $\measuredangle BAD$ is:

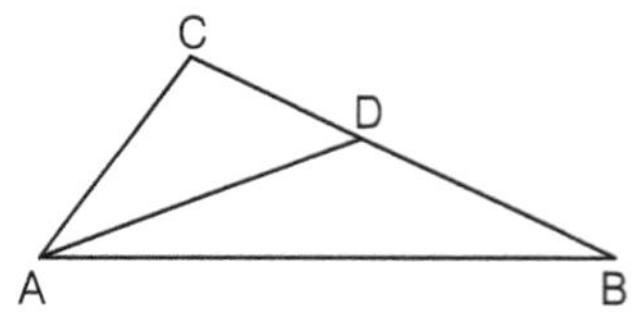

(A) 30° (B) 20°

(C) $22\frac{1}{2}°$ (D) 10°

(E) 15°

45. If two real numbers x and y satisfy the equation $x/y = x - y$, then:

(A) $x \geq 4$ and $x \leq 0$ where $x \geq a$ means that x can take any value greater than a or equal to a

(B) y can equal 1 (C) both x and y must be irrational

(D) x and y cannot both be integers

(E) both x and y must be rational

46. Two perpendicular chords intersect in a circle. The segments of one chord are 3 and 4, the segments of the other are 6 and 2. Then, the diameter of the circle is:

(A) $\sqrt{89}$ (B) $\sqrt{56}$ (C) $\sqrt{61}$

(D) $\sqrt{75}$ (E) $\sqrt{65}$

47. In circle O, the midpoint of radius OX is Q, at Q; $AB \perp XY$. The semi-circle with AB as diameter intersects XY in M. Line AM intersects circle O in C, and line BM intersects circle O in D. Line AD is drawn. Then, if the radius of circle O is r, $\overline{AD}$ is:

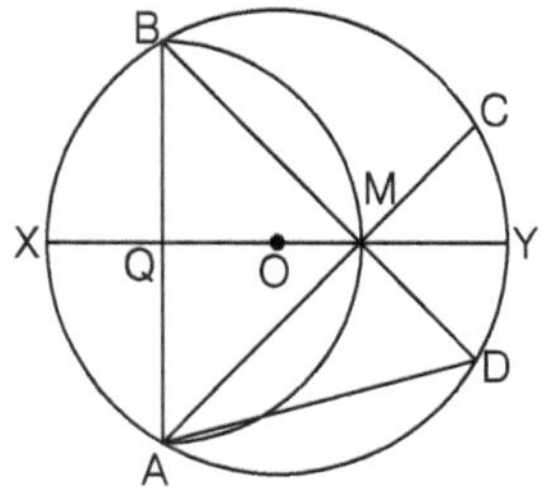

(A) $r\sqrt{2}$ (B) r

(C) not a side of an inscribed regular polygon

(D) $\frac{r\sqrt{3}}{2}$ (E) $r\sqrt{3}$

48. Let ABC be an equilateral triangle inscribed in circle O. M is a point on arc BC. Lines AM, BM, and CM are drawn. Then $\overline{AM}$ is:

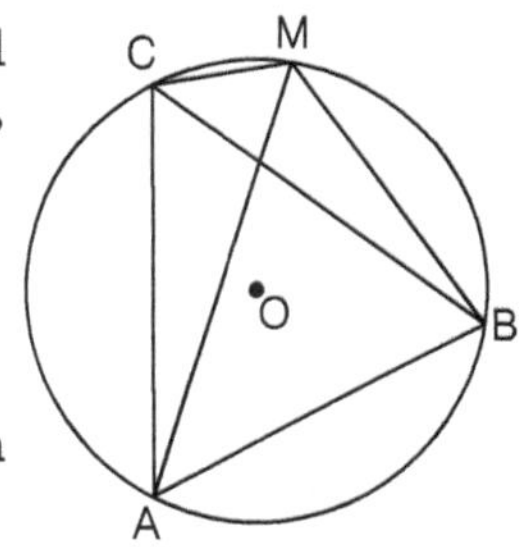

(A) equal to $\overline{BM} + \overline{CM}$

(B) less than $\overline{BM} + \overline{CM}$

(C) greater than $\overline{BM} + \overline{CM}$

(D) equal, less than, or greater than $\overline{BM} + \overline{CM}$, depending upon the position of M

(E) none of these

49. The parallel sides of a trapezoid are 3 and 9. The non-parallel sides are 4 and 6. A line parallel to the bases divides the trapezoid into two trapezoids of equal perimeters. The ratio in which each of the non-parallel sides is divided is:

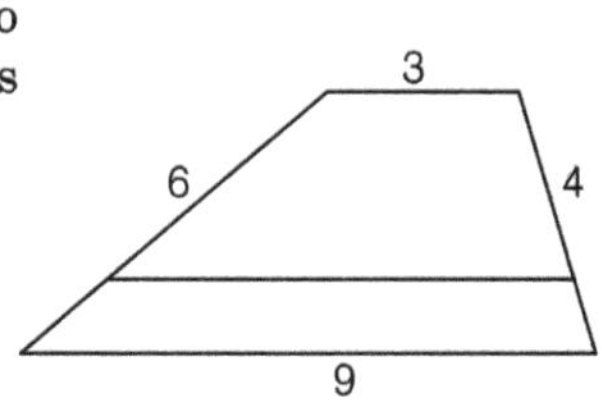

(A) 4 : 3 (B) 3 : 2

(C) 4 : 1 (D) 3 : 1

(E) 6 : 1

50. In circle O, G is a moving point on diameter AB. AA' is drawn perpendicular to AB and equal to AG. BB' is drawn perpendicular to AB, on the same side of diameter AB as AA', and equal to $\overline{BG}$. Let O' be the midpoint of $A'B'$. Then, as G moves from A to B, point O':

(A) moves on a straight line parallel to AB

(B) remains stationary

(C) moves on a straight line perpendicular to AB

(D) moves in a small circle intersecting the given circle

(E) follows a path which is neither a circle nor a straight line

1958 Examination

Part 1

1. The value of $[2-3(2-3)^{-1}]^{-1}$ is:

(A) 5 (B) -5 (C) $\frac{1}{5}$

(D) $-\frac{1}{5}$ (E) $\frac{5}{3}$

2. If $\frac{1}{x}-\frac{1}{y}=\frac{1}{z}$, then z equals:

(A) $y-x$ (B) $x-y$ (C) $\frac{y-x}{xy}$

(D) $\frac{xy}{y-x}$ (E) $\frac{xy}{x-y}$

3. Of the following expressions the one equal to $\frac{a^{-1}b^{-1}}{a^{-3}-b^{-3}}$ is:

(A) $\frac{a^2b^2}{b^2-a^2}$ (B) $\frac{a^2b^2}{b^3-a^3}$ (C) $\frac{ab}{b^3-a^3}$

(D) $\frac{a^3-b^3}{ab}$ (E) $\frac{a^3b^3}{a-b}$

4. In the expression $\frac{x+1}{x-1}$ each x is replaced by $\frac{x+1}{x-1}$. The resulting expression, evaluated for $x=\frac{1}{2}$, equals:

(A) 3 (B) -3 (C) 1

(D) -1 (E) none of these

5. The expression $2+\sqrt{2}+\frac{1}{2+\sqrt{2}}+\frac{1}{\sqrt{2}-2}$ equals:

(A) 2 (B) $2-\sqrt{2}$ (C) $2+\sqrt{2}$

(D) $2\sqrt{2}$ (E) $\frac{\sqrt{2}}{2}$

6. The arithmetic mean between $\frac{x+a}{x}$ and $\frac{x-a}{x}$, when $x \neq 0$, is (the symbol $\neq$ means "not equal to"):

(A) 2, if $a \neq 0$ (B) 1 (C) 1, only if $a = 0$

(D) a/x (E) x

7. A straight line joins the points $(-1, 1)$ and $(3, 9)$. Its x-intercept is:

(A) $\frac{-3}{2}$ (B) $\frac{-2}{3}$ (C) $\frac{2}{5}$

(D) 2 (E) 3

8. Which of these four numbers $\sqrt{\pi^2}$, $\sqrt[3]{.8}$, $\sqrt[4]{.00016}$, $\sqrt[3]{-1}\cdot\sqrt{(.09)^{-1}}$, is (are) rational:

(A) none (B) all (C) the first and fourth

(D) only the fourth (E) only the first

9. A value of x satisfying the equation $x^2 + b^2 = (a - x)^2$ is:

(A) $\frac{b^2+a^2}{2a}$ (B) $\frac{b^2-a^2}{2a}$ (C) $\frac{a^2-b^2}{2a}$

(D) $\frac{a-b}{2}$ (E) $\frac{a^2-b^2}{2}$

10. For what real values of k, other than $k = 0$, does the equation $x^2 + kx + k^2 = 0$ have real roots? (The symbol $x \geq a$ means that x can take on all values greater than a and the value a itself; $x \leq a$ has the corresponding meaning with "less than").

(A) $k < 0$ (B) $k > 0$ (C) $k \geq 1$

(D) all values of k (E) no values of k

11. The number of roots satisfying the equation $\sqrt{5-x} = x\sqrt{5-x}$ is:

(A) unlimited (B) 3 (C) 2

(D) 1 (E) 0

12. If $P = \frac{s}{(1+k)^n}$ then n equals:

(A) $\frac{\log(s/P)}{\log(1+k)}$ (B) $\log\frac{s}{P(1+k)}$ (C) $\log\frac{s-P}{1+k}$

(D) $\log\frac{s}{P} + \log(1+k)$ (E) $\frac{\log s}{\log P(1+k)}$

13. The sum of two numbers is 10, their product is 20. The sum of their reciprocals is:

(A) $\frac{1}{10}$ (B) $\frac{1}{2}$ (C) 1

(D) 2 (E) 4

14. At a dance party a group of boys and girls exchange dances as follows: one boy dances with 5 girls, a second boy dances with 6 girls, and so on, the last boy dancing with all the girls. If b represents the number of boys and g the number of girls, then:

(A) $b = g$ (B) $b = g/5$ (C) $b = g - 4$
(D) $b = g - 5$
(E) It is impossible to determine a relation between b and g without knowing the total number of boys and girls.

15. A quadrilateral is inscribed in a circle. If an angle is inscribed into each of the four segments outside the quadrilateral, the sum of these four angles, expressed in degrees, is:

(A) 1080 (B) 900 (C) 720
(D) 540 (E) 360

16. The area of a circle inscribed in a regular hexagon is 100π. The area of the hexagon is:

(A) 600 (B) 300 (C) $200\sqrt{2}$
(D) $200\sqrt{3}$ (E) $120\sqrt{5}$

17. If x is positive and $\log x \geq \log 2 + \frac{1}{2}\log x$, then:

(A) x has no minimum or maximum value
(B) the maximum value of x is 1
(C) the minimum value of x is 1
(D) the maximum value of x is 4
(E) the minimum value of x is 4

18. The area of a circle is doubled when its radius r is increased by n. Then r equals:

(A) $n(\sqrt{2} + 1)$ (B) $n(\sqrt{2} - 1)$ (C) n
(D) $n(2 - \sqrt{2})$ (E) $\dfrac{n\pi}{\sqrt{2} + 1}$

19. The sides of a right triangle are a and b and the hypotenuse is c. A perpendicular from the vertex divides c into segments r and s, adjacent respectively to a and b. If $a : b = 1 : 3$, then the ratio of r to s is:

(A) $1 : 3$ (B) $1 : 9$ (C) $1 : 10$
(D) $3 : 10$ (E) $1 : \sqrt{10}$

20. If $4^x - 4^{x-1} = 24$, then $(2x)^x$ equals:

(A) $5\sqrt{5}$ (B) $\sqrt{5}$ (C) $25\sqrt{5}$
(D) 125 (E) 25

Part 2

21. In the accompanying figure CE and DE are equal chords of a circle with center O. Arc AB is a quarter-circle. Then, the ratio of the area of triangle CED to the area of triangle AOB is

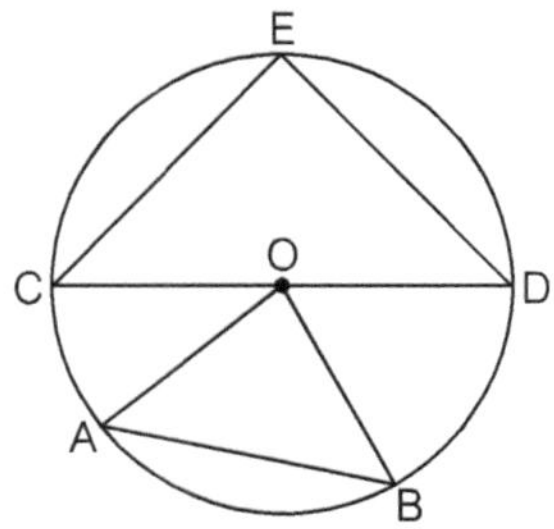

(A) $\sqrt{2}:1$ (B) $\sqrt{3}:1$ (C) $4:1$
(D) $3:1$ (E) $2:1$

22. A particle is placed on the parabola $y = x^2 - x - 6$ at a point P whose ordinate is 6. It is allowed to roll along the parabola until it reaches the nearest point Q whose ordinate is -6. The horizontal distance traveled by the particle (the numerical value of the difference in the abscissas of P and Q) is:

(A) 5 (B) 4 (C) 3
(D) 2 (E) 1

23. If, in the expression $x^2 - 3$, x increases or decreases by a positive amount a, the expression changes by an amount:

(A) $\pm 2ax + a^2$ (B) $2ax \pm a^2$
(C) $\pm a^2 - 3$ (D) $(x + a)^2 - 3$
(E) $(x - a)^2 - 3$

24. A man travels m feet due north at *2 minutes per mile*. He returns due south to his starting point at *2 miles per minute*. The average rate in miles per hour for the entire trip is:

(A) 75 (B) 48 (C) 45
(D) 24
(E) impossible to determine without knowing the value of m

25. If $\log_k x \cdot \log_5 k = 3$, then x equals:

(A) k^5 (B) $5k^3$
(C) k^3 (D) 243
(E) 125

26. A set of n numbers has the sum s. Each number of the set is increased by 20, then multiplied by 5, and then decreased by 20. The sum of the numbers in the new set thus obtained is:

(A) $s + 20n$ (B) $5s + 80n$ (C) s
(D) $5s$ (E) $5s + 4n$

27. The points $(2, -3)$, $(4, 3)$ and $(5, k/2)$ are on the same straight line. The value (s) of k is (are):

(A) 12 (B) -12 (C) ± 12
(D) 12 or 6 (E) 6 or $6\frac{2}{3}$

28. A 16-quart radiator is filled with water. Four quarts are removed and replaced with pure antifreeze liquid. Then four quarts of the mixture are removed and replaced with pure antifreeze. This is done a third and a fourth time. The fractional part of the final mixture that is water is:

(A) $\frac{1}{4}$ (B) $\frac{81}{256}$ (C) $\frac{27}{64}$
(D) $\frac{37}{64}$ (E) $\frac{175}{256}$

29. In a general triangle ADE (as shown) lines EB and EC are drawn. Which of the following angle relations is true?

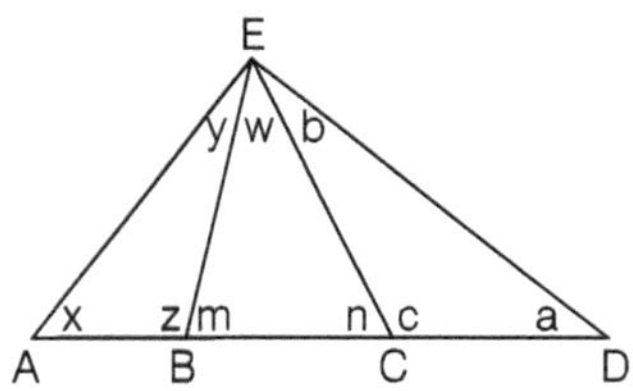

(A) $x + z = a + b$ (B) $y + z = a + b$ (C) $m + x = w + n$
(D) $x + z + n = w + c + m$
(E) $x + y + n = a + b + m$

30. If $xy = b$ and $\frac{1}{x^2} + \frac{1}{y^2} = a$, then $(x + y)^2$ equals:

(A) $(a + 2b)^2$ (B) $a^2 + b^2$ (C) $b(ab + 2)$
(D) $ab(b + 2)$ (E) $\frac{1}{a} + 2b$

31. The altitude drawn to the base of an isosceles triangle is 8, and the perimeter is 32. The area of the triangle is

(A) 56 (B) 48 (C) 40
(D) 32 (E) 24

32. With \$1000 a rancher is to buy steers at \$25 each and cows at \$26 each. If the number of steers s and the number of cows c are both positive integers, then:

(A) this problem has no solution

(B) there are two solutions with s exceeding c

(C) there are two solutions with c exceeding s

(D) there is one solution with s exceeding c

(E) there is one solution with c exceeding s

33. For one root of $ax^2 + bx + c = 0$ to be double the other, the coefficients a, b, c must be related as follows:

(A) $4b^2 = 9c$ (B) $2b^2 = 9ac$ (C) $2b^2 = 9a$

(D) $b^2 - 8ac = 0$ (E) $9b^2 = 2ac$

34. The numerator of a fraction is $6x + 1$, the denominator is $7 - 4x$, and x can have any value between -2 and 2, both included. The values of x for which the numerator is greater than the denominator are:

(A) $\frac{3}{5} < x \le 2$ (B) $\frac{3}{5} \le x \le 2$ (C) $0 < x \le 2$

(D) $0 \le x \le 2$ (E) $-2 \le x \le 2$

35. A triangle is formed by joining three points whose coordinates are integers. If the x-unit and the y-unit are each 1 inch, then the area of the triangle, in square inches:

(A) must be an integer (B) may be irrational (C) must be irrational

(D) must be rational

(E) will be an integer only if the triangle is equilateral.

36. The sides of a triangle are 30, 70, and 80 units. If an altitude is dropped upon the side of length 80, the larger segment cut off on this side is:

(A) 62 (B) 63 (C) 64

(D) 65 (E) 66

37. The first term of an arithmetic series of consecutive integers is $k^2 + 1$. The sum of $2k + 1$ terms of this series may be expressed as:

(A) $k^3 + (k+1)^3$ (B) $(k-1)^3 + k^3$ (C) $(k+1)^3$

(D) $(k+1)^2$ (E) $(2k+1)(k+1)^2$

38. Let r be the distance from the origin to a point P with coordinates x and y. Designate the ratio y/r by s and ratio x/r by c. Then the values of $s^2 - c^2$ are limited to the numbers:

(A) less than -1 and greater than $+1$, both excluded

(B) less than -1 and greater than $+1$, both included

(C) between -1 and $+1$, both excluded

(D) between -1 and $+1$, both included

(E) -1 and $+1$ only

39. The symbol $|x|$ means x if x is not negative and $-x$ if x is not positive. We may then say concerning the solution of
$|x|^2 + |x| - 6 = 0$
that:
(A) there is only one root
(B) the sum of the roots is $+1$
(C) the sum of the roots is 0
(D) the product of the roots is $+4$
(E) the product of the roots is -6

40. Given $a_0 = 1$, $a_1 = 3$, and the general relation $a_n^2 - a_{n-1}a_{n+1} = (-1)^n$ for $n \geq 1$. Then a_3 equals:
(A) $\frac{13}{27}$ (B) 33 (C) 21
(D) 10 (E) -17

Part 3

41. The roots of $Ax^2 + Bx + C = 0$ are r and s. For the roots of $x^2 + px + q = 0$ to be r^2 and s^2, p must equal:
(A) $\frac{B^2 - 4AC}{A^2}$ (B) $\frac{B^2 - 2AC}{A^2}$ (C) $\frac{2AC - B^2}{A^2}$
(D) $B^2 - 2C$ (E) $2C - B^2$

42. In a circle with center O chord AB = chord AC. Chord AD cuts BC in E. If $\overline{AC} = 12$ and $\overline{AE} = 8$, then $\overline{AD}$ equals:
(A) 27 (B) 24 (C) 21
(D) 20 (E) 18

43. AB is the hypotenuse of a right triangle ABC. Median $\overline{AD} = 7$ and median $\overline{BE} = 4$. Then length of AB is:
(A) 10 (B) $5\sqrt{3}$ (C) $5\sqrt{2}$
(D) $2\sqrt{13}$ (E) $2\sqrt{15}$

44. Given the true statements: (1) If a is greater than b, then c is greater than d (2) If c is less than d, then e is greater than f. A valid conclusion is:
(A) If a is less than b, then e is greater than f
(B) If e is greater than f, then a is less than b
(C) If e is less than f, then a is greater than b
(D) If a is greater than b, then e is less than f
(E) none of these

45. A check is written for x dollars and y cents, x and y both two-digit numbers. In error it is cashed for y dollars and x cents, the incorrect amount exceeding the correct amount \$17.82. Then:

(A) x cannot exceed 70

(B) y can equal $2x$

(C) the amount of the check cannot be a multiple of 5

(D) the incorrect amount can equal twice the correct amount

(E) the sum of the digits of the correct amount is divisible by 9

46. For values of x less than 1 but greater than -4, the expression $\frac{x^2 - 2x + 2}{2x - 2}$ has:

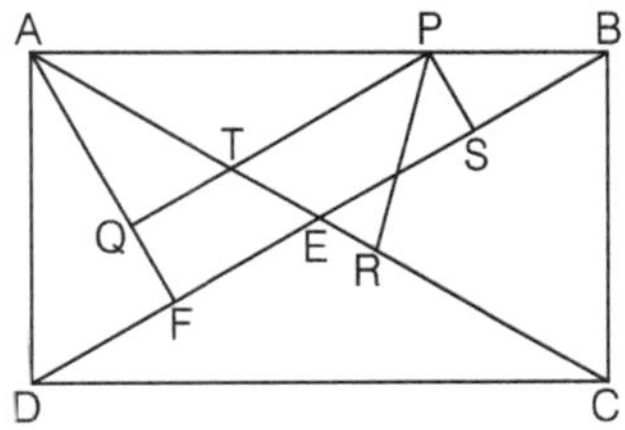

(A) no maximum or minimum value

(B) a minimum value of $+1$

(C) a maximum value of $+1$

(D) a minimum value of -1

(E) a maximum value of -1

47. $ABCD$ is a rectangle (see the accompanying diagram) with P any point on AB. $PS \perp BD$ and $PR \perp AC$. $AF \perp BD$ and $PQ \perp AF$.
Then $\overline{PR} + \overline{PS}$ is equal to:

(A) $\overline{PQ}$ (B) $\overline{AE}$ (C) $\overline{PT} + \overline{AT}$

(D) $\overline{AF}$ (E) $\overline{EF}$

48. Diameter AB of a circle with centre O is 10 units. C is a point 4 units from A, and on AB. D is a point 4 units from B, and on AB. P is any point on the circle. Then the broken-line path from C to P to D:

(A) has the same length for all positions of P

(B) exceeds 10 units for all positions of P

(C) cannot exceed 10 units

(D) is shortest when CPD is a right triangle

(E) is longest when P is equidistant from C and D.

49. In the expansion of $(a + b)^n$ there are $n + 1$ dissimilar terms. The number of dissimilar terms in the expansion of $(a + b + c)^{10}$ is:

(A) 11 (B) 33 (C) 55
(D) 66 (E) 132

50. In this diagram a scheme is indicated for associating all the points of segment AB with those of segment $A'B'$, and reciprocally. To describe this association scheme analytically let x be the distance from a point P on AB to D and let y be the distance from the associated point P' of $A'B'$ to D'. Then for any pair of associated points, if $x = a$, $x + y$ equals:

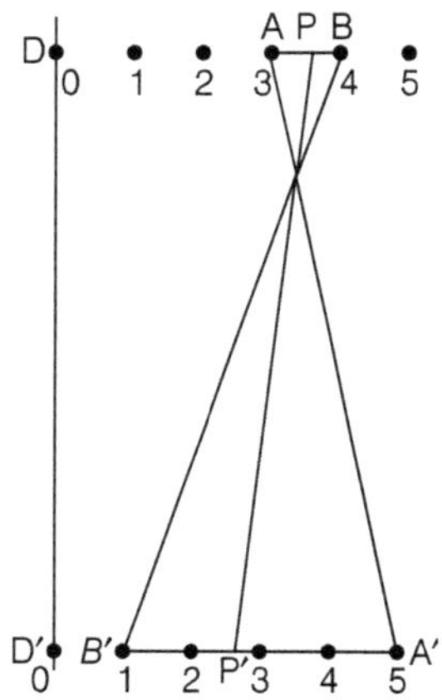

(A) $13a$ (B) $17a - 51$ (C) $17 - 3a$
(D) $\frac{17 - 3a}{4}$ (E) $12a - 34$

1959 Examination

Part 1

1. Each edge of cube is increased by 50%. The percent of increase in the surface area of the cube is

   (A) 50 (B) 125 (C) 150
   (D) 300 (E) 750

2. Through a point P inside the triangle ABC a line is drawn parallel to the base AB, dividing the triangle into two equal areas. If the altitude to AB has a length of 1, then the distance from P to AB is:

   (A) $\frac{1}{2}$ (B) $\frac{1}{4}$ (C) $2-\sqrt{2}$
   (D) $\frac{2-\sqrt{2}}{2}$ (E) $\frac{2+\sqrt{2}}{8}$

3. If the diagonals of a quadrilateral are perpendicular to each other, the figure would always be included under the general classification:

   (A) rhombus (B) rectangle (C) square
   (D) isosceles trapezoid (E) none of these

4. If 78 is divided into three parts which are proportional to $1, \frac{1}{3}, \frac{1}{6}$, the middle part is:

   (A) $9\frac{1}{3}$ (B) 13 (C) $17\frac{1}{3}$
   (D) $18\frac{1}{3}$ (E) 26

5. The value of $(256)^{.16} \cdot (256)^{.09}$ is:

   (A) 4 (B) 16 (C) 64
   (D) 256.25 (E) -16

6. Given the true statement : If a quadrilateral is a square, then it is a rectangle. It follows that, of the converse and the inverse of this true statement:

   (A) only the converse is true (B) only the inverse is true
   (C) both are true (D) neither is true
   (E) the inverse is true, but the converse is sometimes true

7. The sides of a right triangle are a, $a+d$, and $a+2d$, with a and d both positive. The ratio of a to d is:

(A) $1:3$ (B) $1:4$ (C) $2:1$
(D) $3:1$ (E) $3:4$

8. The value of $x^2-6x+13$ can never be less than:

(A) 4 (B) 4.5 (C) 5
(D) 7 (E) 13

9. A farmer divides his herd of n cows among his four sons so that one son gets one-half the herd, a second son, one-fourth, a third son, one-fifth, and the fourth son, 7 cows. Then n is:

(A) 80 (B) 100 (C) 140
(D) 180 (E) 240

10. In triangle ABC, with $\overline{AB}=\overline{AC}=3.6$, a point D is taken on AB at a distance 1.2 from A. Point D is joined to point E in the prolongation of AC so that triangle AED is equal in area to triangle ABC. Then $\overline{AE}$ equals:

(A) 4.8 (B) 5.4 (C) 7.2
(D) 10.8 (E) 12.6

11. The logarithm of .0625 to the base 2 is:

(A) .025 (B) .25 (C) 5
(D) -4 (E) -2

12. By adding the same constant to each of 20, 50, 100 a geometric progression results. The common ratio is:

(A) $\frac{5}{3}$ (B) $\frac{4}{3}$ (C) $\frac{3}{2}$
(D) $\frac{1}{2}$ (E) $\frac{1}{3}$

13. The arithmetic mean (average) of a set of 50 numbers is 38. If two numbers, namely, 45 and 55, are discarded, the mean of the remaining set of numbers is:

(A) 36.5 (B) 37 (C) 37.2
(D) 37.5 (E) 37.52

14. Given the set S whose elements are zero and the even integers, positive and negative. Of the five operations applied to any pair of elements: (1) addition (2) subtraction (3) multiplication (4) division (5) finding the arithmetic mean (average), those operations that yield only elements of S are:

(A) all (B) 1, 2, 3, 4 (C) 1, 2, 3, 5
(D) 1, 2, 3 (E) 1, 3, 5

15. In a right triangle the square of the hypotenuse is equal to twice the product of the legs. One of the acute angles of the triangle is:

(A) 15° (B) 30° (C) 45°
(D) 60° (E) 75°

16. The expression $\frac{x^2-3x+2}{x^2-5x+6} \div \frac{x^2-5x+4}{x^2-7x+12}$, when simplified, is:

(A) $\frac{(x-1)(x-6)}{(x-3)(x-4)}$ (B) $\frac{x+3}{x-3}$ (C) $\frac{x+1}{x-1}$
(D) 1 (E) 2

17. If $y = a + \frac{b}{x}$, where a and b are constants, and if $y = 1$ when $x = -1$, and $y = 5$ when $x = -5$, then $a + b$ equals:

(A) -1 (B) 0 (C) 1
(D) 10 (E) 11

18. The arithmetic mean (average) of the first n positive integers is:

(A) $\frac{n}{2}$ (B) $\frac{n^2}{2}$ (C) n
(D) $\frac{n-1}{2}$ (E) $\frac{n+1}{2}$

19. With the use of three different weights, namely, 1 lb., 3 lb., and 9 lb., how many objects of different weights can be weighed, if the objects to be weighed and the given weights may be placed in either pan of the scale?

(A) 15 (B) 13 (C) 11
(D) 9 (E) 7

20. It is given that x varies directly as y and inversely as the square of z, and that $x = 10$ when $y = 4$ and $z = 14$. Then, when $y = 16$ and $z = 7$, x equals:

(A) 180 (B) 160 (C) 154
(D) 140 (E) 120

Part 2

21. If p is the perimeter of an equilateral triangle inscribed in a circle, the area of the circle is:

(A) $\frac{\pi p^2}{3}$ (B) $\frac{\pi p^2}{9}$ (C) $\frac{\pi p^2}{27}$
(D) $\frac{\pi p^2}{81}$ (E) $\frac{\pi p^2\sqrt{3}}{27}$

22. The line joining the midpoints of the diagonals of a trapezoid has length 3. If the longer base is 97, then the shorter base is:

(A) 94 (B) 92 (C) 91
(D) 90 (E) 89

23. The set of solutions for the equation $\log_{10}(a^2 - 15a) = 2$ consists of:

(A) two integers
(B) one integer and one fraction
(C) two irrational numbers
(D) two non-real numbers
(E) no numbers, that is, the set is empty.

24. A chemist has m ounces of salt water that is $m\%$ salt. How many ounces of salt must he add to make a solution that is $2m\%$ salt?

(A) $\dfrac{m}{100+m}$ (B) $\dfrac{2m}{100-2m}$
(C) $\dfrac{m^2}{100-2m}$ (D) $\dfrac{m^2}{100+2m}$
(E) $\dfrac{2m}{100+m}$

25. The symbol $|a|$ means $+a$ if a is greater than or equal to zero, and $-a$ if a is less than or equal to zero, the symbol $<$ means "less than"; the symbol $>$ means "greater than."

The set of values x satisfying the inequality $|3-x|<4$ consists of all x such that:

(A) $x^2 < 49$ (B) $x^2 > 1$ (C) $1 < x^2 < 49$
(D) $-1 < x < 7$ (E) $-7 < x < 1$

26. The base of an isosceles triangle is $\sqrt{2}$. The medians to the legs intersect each other at right angles. The area of the triangle is:

(A) 1.5 (B) 2 (C) 2.5
(D) 3.5 (E) 4

27. Which one of the following statements is *not* true for the equation

$$ix^2 - x + 2i = 0,$$
where $i \equiv \sqrt{-1}$?

(A) The sum of the roots is 2
(B) The discriminant is 9
(C) The roots are imaginary
(D) The roots can be found by using the quadratic formula
(E) The roots can be found by factoring, using imaginary numbers

28. In triangle ABC, AL bisects angle A and CM bisects angle C. Points L and M are on BC and AB, respectively. The sides of triangle ABC are a, b, and c. Then $\frac{\overline{AM}}{\overline{MB}} = k\frac{\overline{CL}}{\overline{LB}}$ where k is:

(A) 1 (B) $\frac{bc}{a^2}$ (C) $\frac{a^2}{bc}$

(D) $\frac{c}{b}$ (E) $\frac{c}{a}$

29. On an examination of n questions a student answers correctly 15 of the first 20. Of the remaining questions he answers one third correctly. All the questions have the same credit. If the student's mark is 50%, how many different values of n can there be?

(A) 4 (B) 3 (C) 2

(D) 1 (E) the problem cannot be solved

30. A can run around a circular track in 40 seconds. B, running in the opposite direction, meets A every 15 seconds. What is B's time to run around the track, expressed in seconds?

(A) $12\frac{1}{2}$ (B) 24 (C) 25

(D) $27\frac{1}{2}$ (E) 55

31. A square, with an area of 40, is inscribed in a semicircle. The area of a square that could be inscribed in the entire circle with the same radius, is:

(A) 80 (B) 100 (C) 120

(D) 160 (E) 200

32. The length l of a tangent, drawn from a point A to a circle, is $\frac{4}{3}$ of the radius r. The (shortest) distance from A to the circle is:

(A) $\frac{1}{2}r$ (B) r (C) $\frac{1}{2}l$

(D) $\frac{2}{3}l$ (E) a value between r and l.

33. A harmonic progression is a sequence of numbers such that their reciprocals are in arithmetic progression.

Let S_n represent the sum of the first n terms of the harmonic progression, for example, S_3 represents the sum of the first three terms. If the first three terms of a harmonic progression are 3, 4, 6, then:

(A) $S_4 = 20$ (B) $S_4 = 25$ (C) $S_5 = 49$

(D) $S_6 = 49$ (E) $S_2 = \frac{1}{2}S_4$

34. Let the roots of $x^2 - 3x + 1 = 0$ be r and s. Then the expression $r^2 + s^2$ is:

(A) a positive integer (B) a positive fraction greater than 1
(C) a positive fraction less than 1
(D) an irrational number
(E) an imaginary number

35. The symbol $\geq$ means "greater than or equal to", the symbol $\leq$ means "less than or equal to".

In the equation $(x - m)^2 - (x - n)^2 = (m - n)^2$, m is a fixed positive number, and n is a fixed negative number. The set of values x satisfying the equation is:

(A) $x \geq 0$ (B) $x \leq n$ (C) $x = 0$
(D) the set of all real numbers
(E) none of these

36. The base of a triangle is 80, and one of the base angles is 60°. The sum of the lengths of the other two sides is 90. The shortest side is:

(A) 45 (B) 40 (C) 36
(D) 17 (E) 12

37. When simplified the product $\left(1 - \frac{1}{3}\right)\left(1 - \frac{1}{4}\right)\left(1 - \frac{1}{5}\right) \ldots (1 - 1/n)$ becomes:

(A) $\frac{1}{n}$ (B) $\frac{2}{n}$ (C) $\frac{2(n-1)}{n}$
(D) $\frac{2}{n(n+1)}$ (E) $\frac{3}{n(n+1)}$

38. If $4x + \sqrt{2x} = 1$, then x:

(A) is an integer (B) is fractional (C) is irrational
(D) is imaginary (E) may have two different values

39. Let S be the sum of the first nine term of the sequence, $x + a, x^2 + 2a, x^3 + 3a, \ldots$. Then S equals:

(A) $\frac{50a + x + x^8}{x+1}$ (B) $50a - \frac{x + x^{10}}{x-1}$ (C) $\frac{x^9 - 1}{x+1} + 45a$
(D) $\frac{x^{10} - x}{x-1} + 45a$ (E) $\frac{x^{11} - x}{x-1} + 45a$

40. In triangle ABC, BD is a median. CF intersects BD at E so that $\overline{BE} = \overline{ED}$. Point F is on AB. Then, if $\overline{BF} = 5$, $\overline{BA}$ equals:

(A) 10 (B) 12 (C) 15
(D) 20 (E) none of these

Part 3

41. On the same side of a straight line three circles are drawn as follows : a circle with a radius of 4 inches is tangent to the line, the other two circles are equal, and each is tangent to the line and to the other two circles. The radius of the equal circles is:

(A) 24 (B) 20 (C) 18
(D) 16 (E) 12

42. Given three positive integers a, b and c. Their greatest common divisor is D; their least common multiple is M. Then, which two of the following statements are true?

(1) the product MD cannot be less than abc

(2) the product MD cannot be greater than abc

(3) MD equals abc if and only if a, b, c are each prime

(4) MD equals abc if and only if a, b, c are relatively prime in pairs (This means : no two have a common factor greater than 1.)

(A) 1, 2 (B) 1, 3 (C) 1, 4
(D) 2, 3 (E) 2, 4

43. The sides of a triangle are 25, 39, and 40. The diameter of the circumscribed circle is:

(A) $\frac{133}{3}$ (B) $\frac{125}{3}$ (C) 42
(D) 41 (E) 40

44. The roots of $x^2 + bx + c = 0$ are both real and greater than 1. Let $s = b + c + 1$. Then, s:

(A) may be less than zero (B) may be equal to zero
(C) must be greater than zero (D) must be less than zero
(E) must be between -1 and $+1$,

45. If $(\log_3 x)(\log_x 2x)(\log_{2x} y) = \log_x x^2$, then y equals:

(A) $\frac{9}{2}$ (B) 9 (C) 18
(D) 27 (E) 81

46. A student on vacation for d days observed that (1) it rained 7 times, morning or afternoon (2) when it rained in the afternoon, it was clear in the morning (3) there were five clear afternoons (4) there were six clear mornings. Then d equals:

(A) 7 (B) 9 (C) 10
(D) 11 (E) 12

47. Assume that the following three statements are true:

I. All freshmen are human.

II. All students are human.

III. Some students think.

Given the following four statements:

(1) All freshmen are students.

(2) Some humans think.

(3) No freshmen think.

(4) Some humans who think are not students.

Those which are logical consequences of I, II, and III are:

(A) 2 (B) 4 (C) 2, 3

(D) 2, 4 (E) 1, 2

48. Given the polynomial $a_0x^n + a_1x^{n-1} + \dots + a_{n-1}x + a_n$, where n is a positive integer or zero, and a_0 is a positive integer. The remaining a's are integers or zero. Set $h = n + a_0 + |a_1| + |a_2| + \dots + |a_n|$. [See example 25 for the meaning of $|x|$.] The number of polynomials with $h = 3$ is:

(A) 3 (B) 5 (C) 6

(D) 7 (E) 9

49. For the infinite series $1 - \frac{1}{2} - \frac{1}{4} + \frac{1}{8} - \frac{1}{16} - \frac{1}{32} + \frac{1}{64} - \frac{1}{128} - \dots$ let S be the (limiting) sum. Then S equals:

(A) 0 (B) $\frac{2}{7}$ (C) $\frac{6}{7}$

(D) $\frac{9}{32}$ (E) $\frac{27}{32}$

50. A club with x members is organized into four committees in accordance with these two rules:

(1) Each member belongs to two and only two committees.

(2) Each pair of committees has one and only one member in common.

Then x:

(A) cannot be determined

(B) has a single value between 8 and 16

(C) has two values between 8 and 16

(D) has a single value between 4 and 8

(E) has two values between 4 and 8

1960 Examination

Part 1

1. If 2 is a solution (root) of $x^3 + hx + 10 = 0$, then h equals:

(A) 10 (B) 9 (C) 2
(D) -2 (E) -9

2. It takes 5 seconds for a clock to strike 6 o'clock beginning at 6:00 o'clock precisely. If the strikings are uniformly spaced, how long, in seconds, does it take to strike 12 o'clock?

(A) $9\frac{1}{5}$ (B) 10 (C) 11
(D) $14\frac{2}{5}$ (E) none of these

3. Applied to a bill for \$10,000 the difference between a discount of 40% and two successive discounts of 36% and 4%, expressed in dollars, is:

(A) 0 (B) 144 (C) 256
(D) 400 (E) 416

4. Each of two angles of a triangle is 60° and the included side is 4 inches. The area of the triangle, in square inches, is:

(A) $8\sqrt{3}$ (B) 8 (C) $4\sqrt{3}$
(D) 4 (E) $2\sqrt{3}$

5. The number of distinct points common to the graphs of $x^2 + y^2 = 9$ and $y^2 = 9$ is:

(A) infinitely many (B) four (C) two
(D) one (E) none

6. The circumference of a circle is 100 inches. The side of a square inscribed in this circle, expressed in inches, is:

(A) $\frac{25\sqrt{2}}{\pi}$ (B) $\frac{50\sqrt{2}}{\pi}$
(C) $\frac{100}{\pi}$ (D) $\frac{100\sqrt{2}}{\pi}$
(E) $50\sqrt{2}$

7. Circle I passes through the center of, and is tangent to, circle II. The area of circle I is 4 square inches. Then the area of circle II, in square inches, is:

(A) 8 (B) $8\sqrt{2}$ (C) $8\sqrt{\pi}$
(D) 16 (E) $16\sqrt{2}$

8. The number 2.5252525 ... can be written as a fraction. When reduced to lowest terms the sum of the numerator and denominator of this fraction is:

(A) 7 (B) 29 (C) 141
(D) 349 (E) none of these

9. The fraction $\frac{a^2 + b^2 - c^2 + 2ab}{a^2 + c^2 - b^2 + 2ac}$ is (with suitable restrictions on the values of a, b, and c):

(A) irreducible (B) reducible to -1
(C) reducible to a polynomial of three terms
(D) reducible to $\frac{a - b + c}{a + b - c}$ (E) reducible to $\frac{a + b - c}{a - b + c}$

10. Given the following six statements:

(1) All women are good drivers (2) Some women are good drivers
(3) No men are good drivers (4) All men are bad drivers
(5) At least one man is a bad driver
(6) All men are good drivers.

The statement that negates statement (6) is:

(A) (1) (B) (2) (C) (3)
(D) (4) (E) (5)

11. For a given value of k the product of the roots of $x^2 - 3kx + 2k^2 - 1 = 0$ is 7. The roots may be characterised as:

(A) integral and positive (B) integral and negative
(C) rational, but not integral (D) irrational
(E) imaginary

12. The locus of the centres of all circles of given radius a, in the same plane, passing through a fixed point, is:

(A) a point (B) a straight line (C) two straight lines
(D) a circle (E) two circles

13. The polygon(s) formed by $y = 3x + 2$, $y = -3x + 2$, and $y = -2$, is (are):

(A) an equilateral triangle (B) an isosceles triangle
(C) a right triangle (D) a triangle and a trapezoid
(E) a quadrilateral

14. If a and b are real numbers, the equation $3x - 5 + a = bx + 1$ has a unique solution x [the symbol $a \neq 0$ means that a is different from zero]:

(A) for all a and b (B) if $a \neq 2b$ (C) if $a \neq 6$
(D) if $b \neq 0$ (E) if $b \neq 3$

15. Triangle I is equilateral with side A, perimeter P, area K, and circumradius R (radius of the circumscribed circle). Triangle II is equilateral with side a, perimeter p, area k, and circumradius r. If A is different from a, then:

(A) $P : p = R : r$ only sometimes (B) $P : p = R : r$ always
(C) $P : p = K : k$ only sometimes (D) $R : r = K : k$ always
(E) $R : r = K : k$ only sometimes

16. In the numeration system with base 5, counting is as follows : 1, 2, 3, 4, 10, 11, 12, 13, 14, 20, ... The number whose description in the decimal system is 69, when described in the base 5 system, is a number with:

(A) two consecutive digits (B) two non-consecutive digits
(C) three consecutive digits (D) three non-consecutive digits
(E) four digits

17. The formula $N = 8 \cdot 10^8 \cdot x^{-3/2}$ gives, for a certain group, the number of individuals whose income exceeds x dollars. The lowest income, in dollars, of the wealthiest 800 individuals is at least:

(A) 10^4 (B) 10^6 (C) 10^8
(D) 10^{12} (E) 10^{16}

18. The pair of equations $3^{x+y} = 81$ and $81^{x-y} = 3$ has:

(A) no common solution
(B) the solution $x = 2$, $y = 2$
(C) the solution $x = 2\frac{1}{2}$, $y = 1\frac{1}{2}$
(D) a common solution in positive and negative integers
(E) none of these

19. Consider equation I : $x + y + z = 46$ where x, y, and z are positive integers, and equation II : $x + y + z + w = 46$, where x, y, z, and w are positive integers. Then

(A) I can be solved in consecutive integers
(B) I can be solved in consecutive even integers
(C) II can be solved in consecutive integers
(D) II can be solved in consecutive even integers
(E) II can be solved in consecutive odd integers.

20. The coefficient of x^7 in the expansion of $\left(\frac{x^2}{2}-\frac{2}{x}\right)^8$ is:

(A) 56 (B) -56 (C) 14
(D) -14 (E) 0

Part 2

21. The diagonal of square I is $a + b$. The perimeter of square II with *twice* the area of I is:

(A) $(a+b)^2$ (B) $\sqrt{2}(a+b)^2$ (C) $2(a+b)$
(D) $\sqrt{8}(a+b)$ (E) $4(a+b)$

22. The equality $(x+m)^2-(x+n)^2=(m-n)^2$, where m and n are unequal non-zero constants, is satisfied by $x = am + bn$ where:

(A) $a=0$, b has a unique non-zero value
(B) $a=0$, b has two non-zero values
(C) $b=0$, a has a unique non-zero value
(D) $b=0$, a has two non-zero values
(E) a and b each have a unique non-zero value.

23. The radius R of a cylindrical box is 8 inches, the height H is 3 inches. The volume $V=\pi R^2H$ is to be increased by the seeme fixed positive amount when R is increased by x inches as when H is increased by x inches. This condition is satisfied by:

(A) no real value of x
(B) one integral value of x
(C) one rational, but not integral, value of x
(D) one irrational value of x
(E) two real values of x

24. If $\log_{2x} 216 = x$, where x is real, then x is:

(A) A non-square, non-cube integer
(B) a non-square, non-cube, non-integral rational number
(C) an irrational number
(D) a perfect square
(E) a perfect cube

25. Let m and n be any two odd numbers, with n less than m. The largest integer which divides all possible numbers of the form m^2-n^2 is:

(A) 2 (B) 4 (C) 6
(D) 8 (E) 16

26. Find the set of x-values satisfying the inequality $\left|\frac{5-x}{3}\right| < 2$.

[The symbol $|a|$ means $+a$ if a is positive. $-a$ if a is negative, 0 if a is zero. The notation $1 < a < 2$ means that a can have any value between 1 and 2, excluding 1 and 2.]

(A) $1 < x < 11$ (B) $-1 < x < 11$ (C) $x < 11$
(D) $x > 11$ (E) $|x| < 6$

27. Let S be the sum of the interior angles of a polygon P for which each interior angle is $7\frac{1}{2}$ times the exterior angle at the same vertex. Then

(A) $S = 2660°$ and P may be regular
(B) $S = 2660°$ and P is not regular
(C) $S = 2700°$ and P is regular
(D) $S = 2700°$ and P is not regular
(E) $S = 2700°$ and P may or may not be regular

28. The equation $x - \frac{7}{x-3} = 3 - \frac{7}{x-3}$ has:

(A) infinitely many integral roots (B) no root
(C) one integral root (D) two equal integral roots
(E) two equal non-integral roots

29. Five times A's money added to B' money is more than \$51.00. Three times A's money minus B's money is \$21.00. If a represents A's money in dollars and b represents B's money in dollars, then:

(A) $a > 9$, $b > 6$ (B) $a > 9$, $b < 6$
(C) $a > 9$, $b = 6$ (D) $a > 9$, but we can put no bounds on b
(E) $2a = 3b$

30. Given the line $3x + 5y = 15$ and a point on this line equidistant from the coordinate axes. Such a point exists in:

(A) none of the quadrants (B) quadrant I only
(C) quadrants I, II only (D) quadrants I, II, III only
(E) each of the quadrants

Part 3

31. For $x^2 + 2x + 5$ to be a factor of $x^4 + px^2 + q$, the values of p and q must be, respectively:

(A) $-2, 5$ (B) $5, 25$ (C) $10, 20$
(D) $6, 25$ (E) $14, 25$

32. In this figure the center of the circle is O. $AB \perp BC$, $ADOE$ is a straight line, $\overline{AP} = \overline{AD}$, and AB has a length twice the radius. Then:

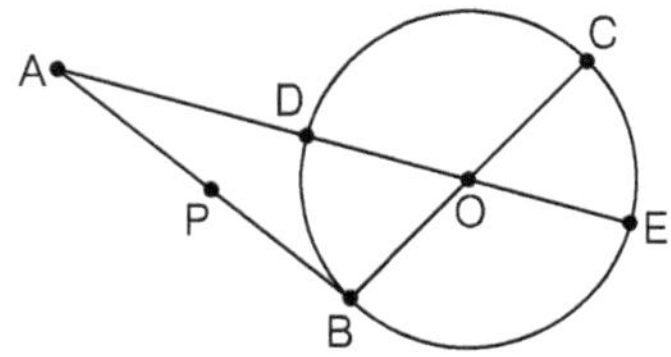

(A) $\overline{AP}^2 = \overline{PB} \cdot \overline{AB}$ (B) $\overline{AP} \cdot \overline{DO} = \overline{PB} \cdot \overline{AD}$ (C) $\overline{AB}^2 = \overline{AD} \cdot \overline{DE}$
(D) $\overline{AB} \cdot \overline{AD} = \overline{OB} \cdot \overline{AO}$ (E) none of these

33. You are given a sequence of 58 terms; each term has the form $P + n$ where P stands for the product $2 \cdot 3 \cdot 5 \cdot \ldots . 61$ of all prime numbers (a prime number is a number divisible only by 1 and itself) less than or equal to 61, and n takes, successively, the values $2, 3, 4, \ldots, 59$. Let N be the number of primes appearing in this sequence. Then, N is:

(A) 0 (B) 16 (C) 17
(D) 57 (E) 58

34. Two swimmers, at opposite ends of a 90-foot pool, start to swim the length of the pool, one at the rate of 3 feet per second, the other at 2 feet per second. They swim back and forth for 12 minutes. Allowing no loss of time at the turns, find the number of times they pass each other.

(A) 24 (B) 21 (C) 20
(D) 19 (E) 18

35. From point P outside a circle, with a circumference of 10 units, a tangent is drawn. Also from P a secant is drawn dividing the circle into unequal arcs with lengths m and n. It is found that t, the length of the tangent, is the mean proportional between m and n. If m and t are integers, then t may have the following number of values:

(A) zero (B) one
(C) two (D) three
(E) infinitely many

36. Let s_1, s_2, s_3 be the respective sums of n, $2n$, $3n$ terms of the same arithmetic progression with a as the first term and d as the common difference. Let $R = s_3 - s_2 - s_1$. Then R is dependent on:

(A) a and d (B) d and n
(C) a and n (D) a, d, and n
(E) neither a nor d nor n

37. The base of a triangle is of length b, and the altitude is of length h. A rectangle of height x is inscribed in the triangle with the base of the rectangle in the base of the triangle. The area of the rectangle is:

(A) $\frac{bx}{h}(h-x)$ (B) $\frac{hx}{b}(b-x)$ (C) $\frac{bx}{h}(h-2x)$

(D) $x(b-x)$ (E) $x(h-x)$

38. In this diagram AB and AC are the equal sides of an isosceles triangle ABC, in which is inscribed equilateral triangle DEF. Designate angle BFD by a, angle ADE by b, and angle FEC by c. Then:

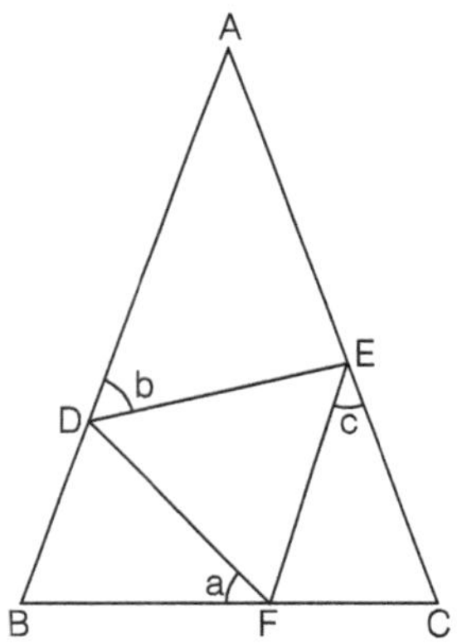

(A) $b = \frac{a+c}{2}$ (B) $b = \frac{a-c}{2}$ (C) $a = \frac{b-c}{2}$

(D) $a = \frac{b+c}{2}$

(E) none of these

39. To satisfy the equation $\frac{a+b}{a} = \frac{b}{a+b}$, a and b must be:

(A) both rational

(B) both real but not rational

(C) both not real

(D) one real, one not real

(E) one real, one not real or both not real

40. Given right triangle ABC with legs $\overline{BC} = 3$, $\overline{AC} = 4$. Find the length of the shorter *angle trisector* from C to the hypotenuse:

(A) $\frac{32\sqrt{3}-24}{13}$ (B) $\frac{12\sqrt{3}-9}{13}$ (C) $6\sqrt{3}-8$

(D) $\frac{5\sqrt{10}}{6}$ (E) $\frac{25}{12}$

ANSWER KEYS

II

ANSWER KEYS

Examinations 1950

1. C	2. D	3. E	4. A	5. A	6. C	7. B	8. C	9. A	10. D
11. B	12. C	13. D	14. D	15. E	16. B	17. C	18. E	19. C	20. D
21. B	22. D	23. B	24. E	25. D	26. E	27. B	28. B	29. B	30. A
31. C	32. D	33. D	34. D	35. B	36. A	37. E	38. B	39. E	40. D
41. D	42. D	43. E	44. C	45. A	46. E	47. C	48. C	49. D	50. E

Examinations 1951

1. B	2. D	3. B	4. D	5. B	6. D	7. B	8. C	9. D	10. C
11. C	12. C	13. C	14. C	15. E	16. C	17. D	18. D	19. E	20. C
21. C	22. B	23. B	24. D	25. A	26. E	27. E	28. C	29. D	30. C
31. D	32. D	33. C	34. A	35. A	36. B	37. D	38. A	39. A	40. C
41. B	42. C	43. E	44. E	45. A	46. A	47. D	48. C	49. D	50. D

Examinations 1952

1. B	2. B	3. A	4. C	5. A	6. E	7. B	8. A	9. D	10. B
11. C	12. E	13. E	14. B	15. C	16. E	17. C	18. D	19. D	20. E
21. B	22. B	23. A	24. B	25. D	26. C	27. E	28. C	29. A	30. A
31. D	32. D	33. B	34. E	35. A	36. E	37. B	38. D	39. A	40. E
41. C	42. D	43. A	44. C	45. E	46. C	47. D	48. A	49. C	50. D

Examinations 1953

1. B	2. D	3. D	4. D	5. C	6. A	7. D	8. C	9. D	10. C
11. C	12. B	13. B	14. E	15. B	16. D	17. B	18. C	19. C	20. D
21. D	22. B	23. B	24. A	25. C	26. A	27. D	28. D	29. D	30. D
31. A	32. D	33. C	34. C	35. E	36. C	37. E	38. C	39. A	40. C
41. E	42. E	43. C	44. A	45. E	46. D	47. D	48. D	49. E	50. B

Examinations 1954

1. E	2. C	3. C	4. E	5. A	6. D	7. B	8. C	9. C	10. C
11. A	12. C	13. B	14. E	15. D	16. D	17. A	18. D	19. D	20. B
21. C	22. B	23. D	24. A	25. D	26. B	27. D	28. B	29. A	30. B
31. A	32. E	33. D	34. D	35. A	36. C	37. B	38. B	39. E	40. C
41. B	42. D	43. B	44. E	45. D	46. E	47. B	48. C	49. C	50. A

Examinations 1955

1. D	2. B	3. B	4. E	5. D	6. B	7. B	8. D	9. D	10. A
11. C	12. D	13. C	14. A	15. D	16. E	17. C	18. A	19. B	20. A
21. B	22. D	23. C	24. E	25. E	26. E	27. B	28. E	29. E	30. B
31. D	32. B	33. A	34. C	35. B	36. E	37. B	38. B	39. B	40. A
41. A	42. C	43. E	44. A	45. A	46. B	47. C	48. B	49. C	50. C

Examinations 1956

1. A	2. D	3. D	4. E	5. C	6. B	7. C	8. B	9. D	10. B
11. A	12. E	13. C	14. B	15. A	16. C	17. D	18. E	19. D	20. C
21. E	22. B	23. C	24. D	25. C	26. A	27. C	28. D	29. D	30. B
31. E	32. B	33. E	34. A	35. B	36. E	37. E	38. D	39. C	40. A
41. C	42. A	43. C	44. B	45. A	46. A	47. D	48. B	49. E	50. B

Examinations 1957

1. B	2. E	3. E	4. C	5. B	6. A	7. E	8. A	9. B	10. C
11. E	12. C	13. C	14. D	15. B	16. D	17. A	18. B	19. A	20. A
21. E	22. A	23. C	24. B	25. B	26. E	27. A	28. B	29. D	30. D
31. B	32. E	33. E	34. C	35. D	36. D	37. C	38. D	39. E	40. D
41. D	42. C	43. B	44. E	45. A	46. E	47. A	48. A	49. C	50. B

Examinations 1958

1. C	2. D	3. B	4. E	5. A	6. B	7. A	8. D	9. C	10. E
11. C	12. A	13. B	14. C	15. D	16. D	17. E	18. A	19. B	20. C
21. E	22. C	23. A	24. B	25. E	26. B	27. A	28. B	29. E	30. C
31. B	32. E	33. B	34. A	35. D	36. D	37. A	38. D	39. C	40. B
41. C	42. E	43. D	44. E	45. B	46. E	47. D	48. E	49. D	50. C

Examinations 1959

1. B	2. D	3. E	4. C	5. A	6. D	7. D	8. A	9. C	10. D
11. D	12. A	13. D	14. D	15. C	16. D	17. E	18. E	19. B	20. B
21. C	22. C	23. A	24. C	25. D	26. A	27. A	28. E	29. D	30. B
31. B	32. C	33. B	34. A	35. E	36. D	37. B	38. B	39. D	40. C
41. D	42. E	43. B	44. C	45. B	46. B	47. A	48. B	49. B	50. D

Examinations 1960

1. E	2. C	3. B	4. C	5. C	6. B	7. D	8. D	9. E	10. E
11. D	12. D	13. B	14. E	15. B	16. C	17. A	18. E	19. C	20. D
21. E	22. A	23. C	24. A	25. D	26. B	27. E	28. B	29. A	30. C
31. D	32. A	33. A	34. C	35. C	36. B	37. A	38. D	39. E	40. A

SOLUTIONS

III
SOLUTIONS

1950 Solutions

Part 1

1. Numbers proportional to 2, 4, 6 are $2x$, $4x$, $6x$, respectively. $2x + 4x + 6x = 64$; $\therefore x = 5\frac{1}{3}$, $2x = 10\frac{2}{3}$.

2. $16 = 8g - 4$; $\therefore g = 2\frac{1}{2}$; $\therefore R = \left(2\frac{1}{2}\right)10 - 4 = 21.$

3. $x^2 - \frac{8x}{4} + \frac{5}{4} = 0$; $\therefore$ $r_1 + r_2 = -\frac{-8}{4} = 2.$

4. $\frac{a^2 - b^2}{ab} - \frac{b(a-b)}{a(b-a)} = \frac{a^2 - b^2}{ab} + \frac{b^2}{ab} = \frac{a}{b}.$

5. Denote the terms in the geometric progression by

$$a_1 = 8,\ a_2 = 8r\ ,\ \dots\ ,\ a_7 = 8r^6 = 5832.$$

$\therefore$ $r^6 = 729$; $\quad\therefore$ $r = 3$ and $a_5 = 8r^4 = 648.$

6. Solve the second equation for x, substitute into the first equation and simplify :

$$x = -\frac{y+3}{2},\ 2\left(-\frac{y+3}{2}\right)^2 + 6\left(-\frac{y+3}{2}\right) + 5y + 1 = 0; \therefore y^2 + 10y - 7 = 0.$$

7. Placing 1 as indicated shifts the given digits to the left, so that t is now the hundreds' digit and u is now the tens' digit.

$\therefore 100t + 10u + 1.$

8. Let r be the original radius; then $2r$ = new radius, πr^2 = original area, $4\pi r^2$ = new area.

Percent increase in area $= \frac{4\pi r^2 - \pi r^2}{\pi r^2}\cdot 100 = 300.$

9. Of all triangles inscribable in a semi-circle, with the diameter as base, the one with the greatest area is the one with the largest altitude (the radius); that is, the isosceles triangle. Thus the area is $\frac{1}{2}\cdot 2r\cdot r = r^2.$

10. $\dfrac{\sqrt{3}-\sqrt{2}}{\sqrt{3}}\cdot\dfrac{\sqrt{3}+\sqrt{2}}{\sqrt{3}+\sqrt{2}}=\dfrac{1}{3+\sqrt{6}}$.

11. $C=\dfrac{en}{R+nr}=\dfrac{e}{(R/n)+r}$. Therefore an increase in n makes the denominator smaller and, consequently, the fraction C larger.

12. It is a theorem in elementary geometry that the sum of the exterior angles of any (convex) polygon is a constant, namely two straight angles.

13. Set each of the factors equal to zero :

$$x=0,\ x-4=0,\ x^2-3x+2=(x-2)\cdot(x-1)=0;$$

$$\therefore \qquad x=0,4,2,1.$$

14. Geometrically, the problem represents a pair of parallel lines, with slope equal to $\frac{2}{3}$, and hence there is no intersection point. Algebraically, if $a_1x+b_1y=c_1$ and $a_2x+b_2y=c_2$,
then, multiplying the first equation by b_2, the second by b_1, and subtracting, we have

$$(a_1b_2-a_2b_1)x=c_1b_2-c_2b_1; \qquad \therefore\ x=\frac{c_1b_2-c_2b_1}{a_1b_2-a_2b_1}.$$

Thus if $a_1b_2-a_2b_1=0$ and $c_1\neq 0$, $c_2\neq 0$, x is undefined and no solution exists.

15. It is understood that we restrict ourselves to rational coefficients and integral powers of x. With this restriction x^2+4 is not factorable.

Part 2

16. The binomial expansion of $(x+y)^n$ has $n+1$ terms. Thus

$[(a+3b)^2(a-3b)^2]^2=(a^2-9b^2)^4$ has $4+1=5$ terms.

17. To obtain a formula directly from the table involves work beyond the indicated scope. Consequently, the answer must be found by testing the choices offered.

(A) is immediately eliminated since, for equally spaced values of x, the values of y are not equally spaced;
$\therefore y$ cannot be a linear function of x.

(B) fails beyond the second set of values.

(C) satisfies the entire table.

(D) fails with the first set of values.

18. Of the four distributive laws given in statements (1), (2), (3) and (5), only (1) holds in the real number field. Statement (4) is not to be confused with $\log(x/y)=\log x-\log y$. $\therefore$(E) is the correct choice.

19. The job requires md man-days, so that one man can do the job in md days. Therefore, $(m + r)$ men can do the job in $md/(m + r)$ days;

or

$$\frac{x}{d} = \frac{m}{m+r}; \qquad \therefore x = \frac{md}{m+r}.$$

20. By ordinary division-323in this instance a long and tedious operation–the answer (D) is found to be correct. By the Remainder Theorem, we have $R = 1^{18} + 1 = 2$. Note : The use of synthetic division shortens the work considerably, but synthetic division itself is generally regarded as beyond the indicated scope of this examination. The Remainder Theorem is likewise beyond the scope of this examination.

21. $lh = 12$, $hw = 8$ and $lw = 6$. Eliminating h we obtain $l = 3w/2$.

Eliminating l, $3w^2/2 = 6$. $\therefore w = 2$, $l = 3$, $h = 4$, and

$$V = lwh = 24;$$

or

$$V^2 = (lwh)^2 = lh \cdot hw \cdot lw = 12 \cdot 8 \cdot 6 = 4^2 \cdot 6^2. \qquad \therefore V = 4 \cdot 6 = 24.$$

22. Let P be the original price. Then a discount of 10% gives a new price $P - .1P = .9P$. Following this by a discount of 20%, we have

$.9P - .2(.9P) = .72P$.

Thus the net discount is $P - .72P = .28P$ or 28%;

or

$$\text{net discount} = d_1 + d_2 - d_1 d_2 = \frac{10}{100} + \frac{20}{100} - \frac{2}{100} = 28\%$$

23. Let R be the yearly rent. Then $\dfrac{5\frac{1}{2}}{100} \cdot 10{,}000 = R - \dfrac{12\frac{1}{2}}{100} R - 325$;

$\therefore R = \$1000$ and $\dfrac{R}{12} \sim \$83.33$.

24. $\sqrt{x-2} = 4 - x$, $x - 2 = 16 - 8x + x^2$, $0 = x^2 - 9x + 18 = (x-6)(x-3)$;

$\therefore x = 6$, $x = 3$.

A check is obtained only with $x = 3$.

25. $\log_5\left(\dfrac{5^3 \cdot 5^4}{5^2}\right) = \log_5 5^5 = 5 \log_5 5 = 5.$

26. $b = \log_{10} m + \log_{10} n = \log_{10} mn$;

$\therefore \qquad mn = 10^b$;

$\therefore \qquad m = 10^b / n$;

or

$$\log_{10} m = \log_{10} 10^b - \log_{10} n = \log_{10}(10^b / n); \quad \therefore m = 10^b / n.$$

27. If a car travels a distance d at rate r_1 and returns the same distance at rate r_2, then

$$\text{average speed} = \frac{\text{total distance}}{\text{total time}} = \frac{2d}{d/r_1 + d/r_2} = \frac{2r_1r_2}{r_1 + r_2};$$

$$\therefore\ x = \frac{2\cdot 30\cdot 40}{70} = \frac{240}{7} \sim 34 \text{ mph.}$$

28. Notice that A and B both travel for the same amount of time. Since A travels 48 miles while B travels 72 miles, we have

$$t = \frac{48}{r} = \frac{72}{r+4}; \qquad \therefore\ r = 8.$$

29. Since the first machine addresses 500 envelopes in 8 minutes, it addresses 500/8 envelopes in 1 minute and $2\cdot 500/8$ in 2 minutes. If the second machine addresses 500 envelopes in x minutes, then it addresses $2\cdot 500/x$ envelopes in 2 minutes. To determine x so that both machines together address 500 envelopes in 2 minutes we use the condition

$$\frac{2}{8}\cdot 500 + \frac{2}{x}\cdot 500 = 500 \quad \text{or} \quad \frac{500}{8} + \frac{500}{x} = 250$$

or $$\frac{1}{8} + \frac{1}{x} = \frac{1}{2}.$$

30. If B is the original number of boys and G the original number of girls, then

$$\frac{B}{G-15} = 2, \quad \frac{B-45}{G-15} = \frac{1}{5};$$

$$\therefore \qquad G = 40.$$

31. Let x be the number of pairs of blue socks, and let y be the price of a pair of blue socks. Then $2y$ is the price of a pair of black socks and $x\cdot 2y + 4\cdot y = \frac{3}{2}\ (4\cdot 2y + x\cdot y)$. Divide by y and solve for x.

$\therefore\ x = 16;\ \therefore\ 4 : x = 4 : 16 = 1 : 4.$

32. Before sliding, the situation is represented by a right triangle with hypotenuse 25, and horizontal arm 7, so that the vertical arm is 24. After sliding, the situation is represented by a right triangle with hypotenuse 25, and vertical arm 20, so that the horizontal arm is 15. $15 - 7 = 8$ so that (D) is the correct choice.

33. The cross section of the pipe is πR^2.

$\therefore$ large pipe/small pipe $= 6^2/1^2 = 36$.

The large pipe has the carrying capacity of 36 small pipes.

34. Since $C = 2\pi r$, letting r_2 be the original radius, r_1 the final radius, we have $25 - 20 = 2\pi r_1 - 2\pi r_2 = 2\pi(r_1 - r_2)$; $\therefore \qquad r_1 - r_2 = 5/2\pi$.

35. For any right triangle we have $a - r + b - r = c$ where r is the radius of the inscribed circle.
$\therefore 2r = a + b - c = 24 + 10 - 26 = 8; \therefore r = 4.$

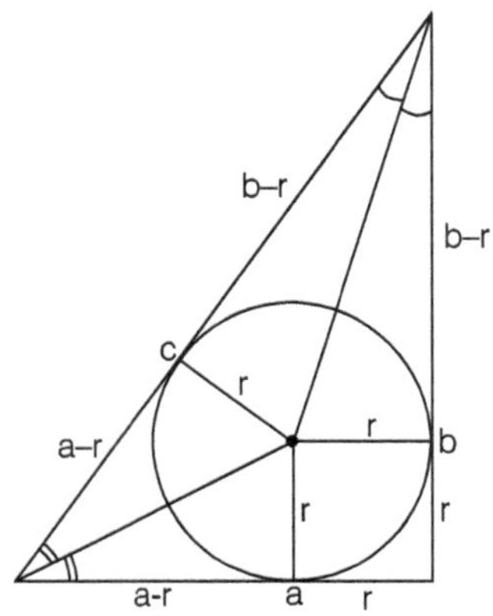

Part 3

36. Let C = merchant's cost, L = list price, M = marked price, S = selling price, and P = profit. Then

$$C = L - \frac{1}{4}L = \frac{3}{4}L, S = M - \frac{1}{5}M = \frac{4}{5}M, S = C + P,$$

$$\frac{4}{5}M = \frac{3}{4}L + \frac{1}{4}\cdot\frac{4}{5}M, \quad M = \frac{5}{3}\cdot\frac{3}{4}L = \frac{5}{4}L.$$

37. Reference to the graph of $y = \log_a x$, $a > 1$, or to the equality $x = a^y$, shows that (A), (B), (C), and (D) are all correct.

38. $\begin{vmatrix} 2x & 1 \\ x & x \end{vmatrix} = 2x\cdot x - 1\cdot x = 3; \therefore x = -1, \frac{3}{2}.$

39. This sequence is an infinite geometric progression with $r = \frac{1}{2} < 1$.

Thus if a is its first term and s its sum, we have

$$s = \frac{a}{1-r} = \frac{2}{1-\frac{1}{2}} = 4.$$

Hence (4) and (5) are correct statements.

40. $\frac{x^2-1}{x-1} = \frac{(x+1)(x-1)}{x-1} = x + 1$, for all $x \neq 1$;

the limit of $x + 1$ as $x \to 1$ is 2, so that (D) is the correct choice.
Note : This problem is beyond the scope of the contest.

41. The graph of the function $y \equiv ax^2 + bx + c, a \neq 0$, is a parabola with its axis of symmetry (the line MV extended) parallel to the y-axis.

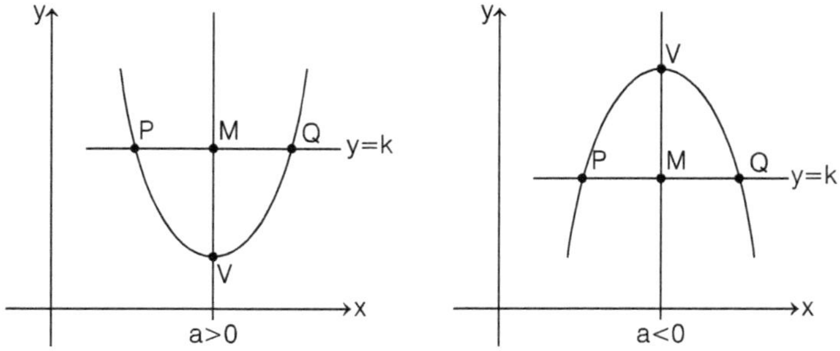

We wish to find the coordinates of V. Let $y = k$ be any line parallel to the x-axis which intersects the parabola at two points, say P and Q. Then the abscissa (the x-coordinate) of the midpoint of the line segment PQ is the abscissa of V. Since P and Q are the intersections of the line $y = k$ with the parabola, the abscissas of P and Q must satisfy the equation.

$ax^2 + bx + c = k$ or $ax^2 + bx + c - k = 0$.

Its roots are :

$$x = -\frac{b}{2a} + \frac{1}{2a}\sqrt{D} \text{ and } x' = -\frac{b}{2a} - \frac{1}{2a}\sqrt{D},$$

where $D = b^2 - 4ac$ is the discriminant.

$$\therefore \text{the abscissa of } V = \frac{x + x'}{2} = -\frac{b}{2a}, \text{ and the}$$

$$\text{ordinate of } V = a\left(-\frac{b}{2a}\right)^2 + b\left(-\frac{b}{2a}\right) + c = -\frac{b^2}{4a} + c = \frac{4ac - b^2}{4a};$$

if $a > 0$, the ordinate of V is the minimum value of the function $ax^2 + bx + c$ and if $a < 0$, it is the maximum value.

42. If $x^{x^{x^{\cdot^{\cdot^{\cdot}}}}} = 2$, then the exponent, which is again $x^{x^{x^{\cdot^{\cdot^{\cdot}}}}}$ is also 2, and we have $x^2 = 2$. Therefore $x = \sqrt{2}$. Note : This problem is beyond the scope of the contest.

43. Rearrange the terms:

$$s_1 = \frac{1}{7} + \frac{1}{7^3} + \ldots = \frac{1/7}{1 - (1/7^2)} = \frac{7}{48};\ s_2 = \frac{2}{7^2} + \frac{2}{7^4} + \ldots = \frac{2/7^2}{1 - (/7^2)} = \frac{2}{48};$$

$$s = s_1 + s_2 = \frac{3}{16}.$$

44. Looking at $y = \log_b x$ or equivalently $x = b^y$, we see that $y = 0$ for $x = 1$. That is, the graph cuts the x-axis at $x = 1$.

45. $d = n(n-3)/2 = 100 \cdot 97/2 = 4850.$

46. In the new triangle $\overline{AB} = \overline{AC} + \overline{BC}$, that is, C lies on the line AB. Consequently, the altitude from C is zero. Therefore, the area of the triangle is zero.

47. By similar triangles, $(h-x)/h = 2x/b$.

$\therefore\ x = bh/(2h+b).$

48. Let P be an arbitrary point in the equilateral triangle ABC with sides of length s, and denote the perpendicular segments by p_a, p_b, p_c. Then

$$\text{Area } ABC = \text{Area } APB + \text{Area } BPC + \text{Area } CPA$$
$$= \frac{1}{2}(sp_a + sp_b + sp_c)$$
$$= \frac{1}{2}s(p_a + p_b + p_c).$$

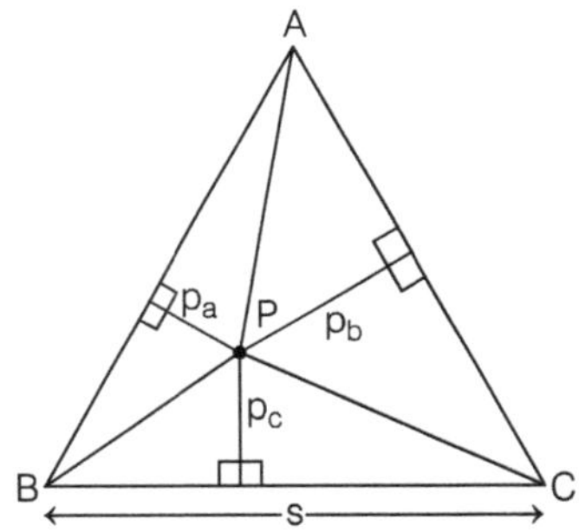

Also, Area $ABC = \frac{1}{2}s \cdot h$, where h is the length of the altitude of ABC.

Therefore, $h = p_a + p_b + p_c$ and this sum does not depend on the location of P.

49. Perhaps the easiest approach to this problem is through coordinate geometry.

Let the triangle be $A(0,0)$, $B(2,0)$ and $C(x, y)$. Then A_1, the midpoint of BC, has the coordinates $[(x+2)/2, y/2]$. Therefore, using the distance formula, we have,

$$\sqrt{\left(\frac{x+2}{2}\right)^2 + \left(\frac{y}{2}\right)^2} = \frac{3}{2}$$

or $$\left(\frac{x+2}{2}\right)^2 + \left(\frac{y}{2}\right)^2 = \frac{9}{4}$$

or $$(x+2)^2 + y^2 = 9.$$

This equation represents a circle with radius 3 inches and center 4 inches from B along BA;

or

as extreme values we may have $a-b=2$ and $a-b=0$, using the notation a, b, c with its usual meaning. From the median formula, with $c=2$ and $m_a=3/2$, we have $2b^2+8=9+a^2$.

$\therefore 2b^2-a^2=1$.

Using the given extreme values in succession, we have $b=5$ and $a=7$, and $b=1$ and $a=1$. Therefore, the maximum distance between the extreme positions of C is 6 inches. These facts, properly collated, lead to the same result as that shown above.

50. After the two-hour chase, at 1:45 P.M. the distance between the two ships is 4 miles. Let t be the number of hours needed for the privateer to overtake the merchantman. Let D and $D+4$ be the distances covered in t hours by the merchantman and the privateer, respectively.

Then Rate of merchantman $t=D$, and Rate of privateer $\cdot\, t=D+4$

or $$\frac{8\cdot 17}{15}t=D+4, \text{ and}$$

$$8t=D;\ \therefore \left(\frac{8\cdot 17}{15}-8\right)t=4.$$

$\therefore t=3\frac{3}{4}$ hours, so that the ships meet at 5:30.

1951 Solutions

Part 1

1. The pedagogic value of this extremely simple problem can be enhanced by a discussion of necessary restrictions on the number-nature of M and N, the permissibility of negative numbers, and so forth.

2. Let the dimensions of the field be w and $2w$. Then $6w = x$, $w = x/6$, and $2w = x/3$. $\therefore\ A = w \cdot 2w = x^2/18$.

3. $A = \frac{1}{2}d^2 = \frac{1}{2}(a + b)^2$.

4. $S = 1(13 \cdot 10) + 2[2(13 \cdot 5) + 2(10 \cdot 5)] = 590$.

5. $S_1 = 10{,}000 + 1000 = 11{,}000$, $S_2 = 11{,}000 - 1100 = 9900$,
$S_1 - S_2 = 1100$. $\therefore$ (B) is the correct choice.

6. If the dimensions are designated by l, w, h, then the bottom, side, and front areas are lw, wh, hl. The product of these areas is $l^2w^2h^2 = (lwh)^2 =$ the square of the volume.

7. Relative error = error/measurement.
Since $.02/10 = .2/100$, the relative errors are the same.

8. Let M be the marked price of the article. $M - .1M = .9M =$ new price. To restore $.9M$ to M requires the addition of $.1M$. $.1M = \frac{1}{9}(.9M)$, and $\frac{1}{9}$, expressed as a percent, is $11\frac{1}{9}\%$.

9. Let P_k be the perimeter of the kth triangle.

Then $P_{k+1} = \frac{1}{2}P_k$.

$$\begin{aligned} S &= P_1 + P_2 + P_3 + \dots \\ &= 3a + \frac{1}{2} \cdot 3a + \frac{1}{2} \cdot \frac{3}{2}a + \dots \\ &= 3a\left(1 + \frac{1}{2} + \frac{1}{4} + \dots\right) = 3a\frac{1}{1 - (1/2)} = 6a. \end{aligned}$$

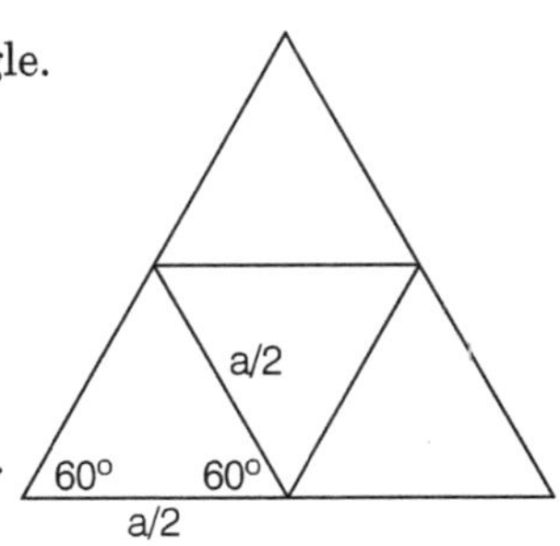

10. (C) is incorrect since doubling the radius of a given circle quadruples the area.

11. The new series is $a^2 + a^2r^2 + a^2r^4 + \ldots$;

$\therefore S = a^2/(1 - r^2)$.

12. In 15 minutes the hour hand moves through an angle of $\frac{1}{4}$ of $30° = 7\frac{1}{2}°$. Therefore, the angle between the hour hand and the minute hand is $22\frac{1}{2}°$.

13. In one day A can do 1/9 of the job. Since B is 50% more efficient than A, B can do (3/2) (1/9) = 1/6 of the job in one day. Therefore, B needs 6 days for the complete job.

14. (C) is incorrect since some terms (primitives) must necessarily remain undefined.

15. $n^3 - n = (n-1)(n)(n+1)$, for integral values of n, represents the product of three consecutive integers. Since in a pair of consecutive integers, there is a multiple of two, and in a triplet of consecutive integers, there is a multiple of three, then $n^3 - n$ is divisible by 6.

Part 2

16. The condition $c = b^2/4a$ implies equal real roots of $f(x) = 0$ for real coefficients, i.e., the curve touches the x-axis at exactly one point and, therefore, the graph is tangent to the x-axis.

17. (A), (B), (C), and (E) are of the form $y = kx$ or $xy = k$, but (D) is not.

18. Let the factors be $Ax + B$ and $Cx + D$. Then

$$(Ax + B)(Cx + D) = ACx^2 + (AD + BC)x + BD$$
$$= 21x^2 + ax + 21.$$

$\therefore AC = 21, BD = 21$. Since 21 is odd, all its factors are odd. Therefore each of the numbers A and C is odd. The same is true for B and D. Since the product of two odd numbers is odd, AD and BC are odd numbers; $a = AD + BC$, the sum of two odd numbers, is even.

19. Such a number can be written as $P\cdot 10^3 + P = P(10^3 + 1) = P(1001)$, where P is the block of three digits that repeats.

$\therefore$ (E) is the correct choice.

20. $$(x+y)^{-1}(x^{-1} + y^{-1}) = \frac{1}{x+y}\left(\frac{1}{x} + \frac{1}{y}\right)$$
$$= \frac{1}{x+y}\cdot\frac{x+y}{xy} = \frac{1}{xy} = x^{-1}y^{-1}$$

21. (C) is not always correct since, if z is negative, $xz < yz$.

22. Since $\log_{10}(a^2 - 15a) = 2$, then $a^2 - 15a = 10^2 = 100$. The solution set for this quadratic equation is $\{20, -5\}$.

23. Since $V = \pi r^2 h$, we must have $\pi(r + x)^2 h = \pi r^2(h + x)$.

$\therefore \quad x = (r^2 - 2rh)/h$.

For $r = 8$ and $h = 3$, we have $x = 5\frac{1}{3}$.

24. $\dfrac{2^{n+4} - 2(2^n)}{2(2^{n+3})}$

$$= \frac{2^n \cdot 2^4 - 2 \cdot 2^n}{2 \cdot 2^n \cdot 2^3} = \frac{2 \cdot 2^n (2^3 - 1)}{2 \cdot 2^n \cdot 2^3} = \frac{7}{8}.$$

25. Let s_1 be the side of the square, a_1 its apothem; then $s_1^2 = 4s_1$ and since $2a_1 = s_1$, we have $4a_1^2 = 8a_1$ or $a_1 = 2$. Let s_2 be the side of the equilateral triangle, h its altitude and a_2 its apothem; then $s_2^2\sqrt{3}/4 = 3s_2$. Since $h = 3a_2$, and $s_2 = 2h/\sqrt{3} = 6a_2/\sqrt{3}$, we have $(36a_2^2/3)\sqrt{3}/4 = 3 \cdot 6a_2/\sqrt{3}$, and $a_2 = 2 = a_1$.

26. After simplification, we have $x^2 - x - m(m + 1) = 0$. For equal roots the discriminate

$$D = 1 + 4m(m + 1) = 0 = (2m + 1)^2.$$

$$\therefore m = -\frac{1}{2}.$$

27. We may eliminate (A) and (D) by proving that corresponding angles are not equal. If (A) is false, (B) is certainly false. We may eliminate (C) by placing the point very close to one side of the triangle. Thus none of the four statements is true.

28. $P = kAV^2, 1 = k \cdot 1 \cdot 16^2; \qquad \therefore k = 1/16^2;$

$36/9 = (1/16^2) \cdot (1) \cdot (V^2); \qquad \therefore V = 32$ (mph).

29. Let the ratio of altitude to base be r. The triangles in the figure satisfy condition (D) but have different shapes.

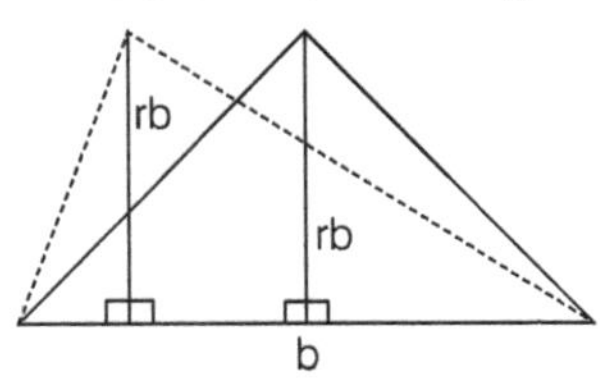

Hence (D) does not determine the shape of a triangle.

30. $\frac{1}{x} = \frac{1}{20} + \frac{1}{80}$; $\therefore$ $x = 16$ (inches);

or

$\frac{20}{100} = \frac{x}{y}$, $y = 5x$

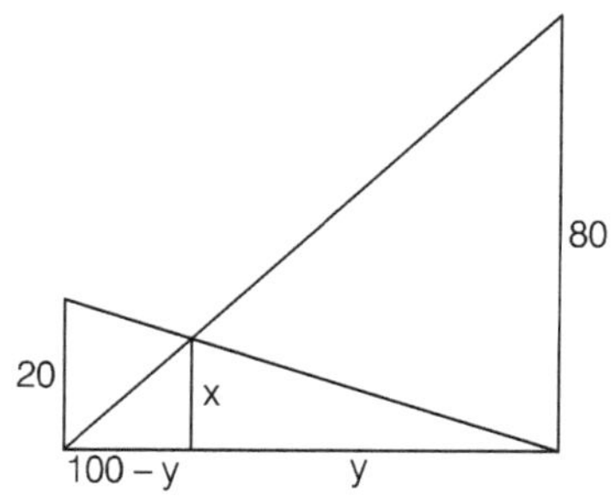

and $\frac{80}{100} = \frac{x}{100 - y} = \frac{x}{100 - 5x}$; $x = 16$ (inches).

31. Let n be the number of people present. Single out a particular individual P. P shakes hands with $(n - 1)$ persons, and this is true for each of the n persons. Since each handshake is between two persons, the total number of handshakes is $n(n-1)/2 = 28$. $\therefore$ $n = 8$;

Or

$$C(n, 2) = n(n-1)/2 = 28.$$

32. The inscribed triangle with maximum perimeter is the isosceles right triangle; in this case $\overline{AC} + \overline{BC} = \overline{AB}\sqrt{2}$. Therefore, in general, $\overline{AC} + \overline{BC} \le \overline{AB}\sqrt{2}$.

33. $(x^2 - 2x = 0) \leftrightarrow (x^2 = 2x) \leftrightarrow (x^2 - 2x + 1 = 1) \leftrightarrow (x^2 - 1 = 2x - 1)$.

$\therefore$(C) is the correct choice.

34. In general, $a^{\log_a N} = N$, with suitable restrictions on a and N;

Or

let $10^{\log_{10} 7} = N$. Taking logarithms of both sides of the equation to the base 10, we have $\log_{10} 7 \cdot \log_{10} 10 = \log_{10} N$. $\therefore N = 7$.

35. Since $c^y = a^z$, $c = a^{z/y}$

$\therefore c^q = a^{(z/y)q} = a^x$; $\therefore x = zq/y$; $\therefore xy = qz$.

Part 3

36. To prove that a geometric figure is a locus it is essential to (1) include all proper points, and (2) exclude all improper points. By this criterion, (B) is insufficient to guarantee condition (1); i.e., (B) does not say that every point satisfying the conditions lies on the locus.

37. Let N denote the number to be found; then we may write the given information as

$$N = 10a_9 + 9 = 9a_8 + 8 = \ldots = 2a_1 + 1.$$

Hence $\qquad N + 1 = 10(a_9 + 1) = 9(a_8 + 1) = \ldots = 2(a_1 + 1),$

i.e., $N + 1$ has factors 2, 3, 4, ... , 10 whose least common multiple is $2^3 \cdot 3^2 \cdot 5 \cdot 7 = 2520$.

$\therefore \qquad N = 2519.$

38. $.03 = 600/x_1$ and $.02 = 600/x_2$; $\therefore x_1 = 20{,}000$ and $x_2 = 30{,}000$;

$\therefore x_2 - x_1 = 10{,}000$ (feet).

39. Since the distance d traveled by the stone and the sound is the same, $d = r_1t_1 = r_2t_2$.

$\therefore$ The times of travel are inversely proportional to the rates of fall : $r_1/r_2 = t_2/t_1$.

$\therefore \dfrac{16t^2/t}{1120} = \dfrac{7.7 - t}{t}$; $\quad \therefore t = 7$ (seconds); $\quad \therefore 16t^2 = 16 \cdot 7^2 = 784$ (feet).

40. Since $x^3 + 1 = (x+1)(x^2 - x + 1)$ and $x^3 - 1 = (x-1)(x^2 + x + 1)$,

the given expression, when simplified, is 1 (with suitable restrictions on the value of x).

41. Since the y-differences are 2, 4, 6, 8, a quadratic function is suggested, thus eliminating (C). A quick check of the other choices shows that (B) is correct;

Or

Make the check with each of the choices given.

42. $x = \sqrt{1 + x}$, $x^2 = 1 + x$,

$x^2 - x - 1 = 0$, and $x \sim 1.62$.

$\therefore \qquad 1 < x < 2.$

43. (E) is incorrect because, when the product of two positive quantities is given, their sum is least when they are equal. To prove this, let one of the numbers be x; then the other number can be written $x + h$ (h may be negative). Then if $x(x + h) = p$, $x = (-h + \sqrt{h^2 + 4p})/2$.

$\therefore x + x + h = \sqrt{h^2 + 4p}$, which has its least value when $h = 0$.

44. Inverting each of the expressions, we have the set of equations:

$\dfrac{1}{y} + \dfrac{1}{x} = \dfrac{1}{a}$, $\dfrac{1}{z} + \dfrac{1}{x} = \dfrac{1}{b}$, $\dfrac{1}{y} + \dfrac{1}{z} = \dfrac{1}{c}$.

$\therefore \dfrac{1}{x} - \dfrac{1}{z} = \dfrac{1}{a} - \dfrac{1}{c}$, $\dfrac{1}{x} + \dfrac{1}{z} = \dfrac{1}{b}$; $\therefore \dfrac{2}{x} = \dfrac{1}{a} + \dfrac{1}{b} - \dfrac{1}{c}$.

$\therefore \qquad x = \dfrac{2abc}{ac + bc - ab}.$

45. $\text{Log}\, 8 = \log 2^3 = 3\log 2$; $\log 9 = \log 3^2 = 2\log 3$; $\log 10 = 1$ and $\log 5 = \log(10/2) = \log 10 - \log 2 = 1 - \log 2$. From these and from the information supplied in the problem,

$\log 2 = \frac{1}{3}\log 8$, $\log 3 = \frac{1}{2}\log 9$, and $\log 5 = 1 - \log 2$

can be found; consequently, so can the logarithm of any number representable as a product of powers of 2, 3 and 5. Since 17 is the only number in the list not representable in this way, (A) is the correct choice.

46. Extend CO so that it meets the circle at E. Since CE is a diameter, $CD \perp DE$. The bisector of angle OCD bisects the subtended arc DE, that is, arc DP = arc PE.

Regardless of the position of C, the corresponding chord DE is always parallel to AB, and hence P always bisects the arc AB.

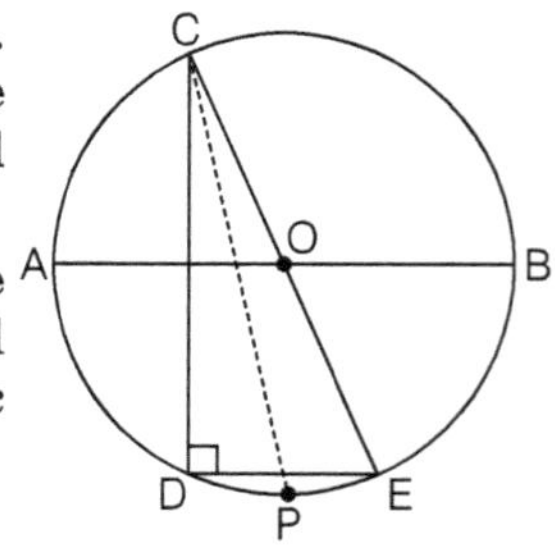

47. Since $\frac{1}{r^2} + \frac{1}{s^2} = \frac{r^2 + s^2}{r^2 s^2} = \frac{(r+s)^2 - 2rs}{(rs)^2}$, and

since $r + s = -\frac{b}{a}$ and $rs = \frac{c}{a}$, we have $\frac{1}{r^2} + \frac{1}{s^2} = \frac{b^2 - 2ac}{c^2}$.

48. The area of the large square is $2a^2$, where a is the radius of the circle. The area of the small square is $4a^2/5$. The ratio is 2 : 5.

49. From the given information it follows that for the right triangle ABC, $(a/2)^2 + b^2 = 25$ and $a^2 + (b/2)^2 = 40$.

$\therefore a^2 = 36$ and $b^2 = 16$.

$\therefore \quad c^2 = a^2 + b^2 = 52;$

$\therefore \quad c = 2\sqrt{13}.$

50. Let t_1, t_2, t_3 be the number of hour, respectively, that the car travels forward, back to pick up Dick, then forward to the destination. Then we may write

$25 \cdot t_1 - 25 \cdot t_2 + 25 \cdot t_3 = 100$ for car

$5 \cdot t_1 + 5 \cdot t_2 + 25 \cdot t_3 = 100$ for Dick

$25 \cdot t_1 + 5 \cdot t_2 + 5 \cdot t_3 = 100$ for Harry.

This system of simultaneous equations is equivalent to the system

$$t_1 - t_2 + t_3 = 4$$
$$t_1 + t_2 + 5t_3 = 20$$
$$5t_1 + t_2 + t_3 = 20$$

whose solution is $t_1 = 3$, $t_2 = 2$, $t_3 = 3$.

Hence $t_1 + t_2 + t_3 = 8$ = total number of hours.

1952 Solutions

Part 1

1. Since $A = \pi r^2$ and π is irrational and r is rational, A is irrational.

   (The product of a rational number and an irrational number is irrational).

2. Average $= \dfrac{20 \cdot 80 + 30 \cdot 70}{20 + 30} = 74.$

3. $a^3 - a^{-3} = a^3 - \dfrac{1}{a^3} = \left(a - \dfrac{1}{a}\right)\left(a^2 + 1 + \dfrac{1}{a^2}\right).$

4. For $P - 1$ pounds the charge is 3 cents per pound. For the first pound the charge is 10 cents.

   $\therefore C = 10 + 3(P - 1);$

   or

   The charge for each pound is 3 cents with an additional charge of 7 cents for the first pound. $\therefore C = 3P + 7 = 3P - 3 + 10 = 10 + 3(P - 1).$

5. $y = mx + b,\ m = \dfrac{y_2 - y_1}{x_2 - x_1} = \dfrac{12 + 6}{6 - 0} = 3; \therefore y = 3x + b.$

   Substituting, $-6 = 3 \cdot 0 + b;$

   $\therefore \qquad b = -6; \therefore y = 3x - 6.$

   This equation is satisfied by (3, 3).

6. The roots are $(7 \pm \sqrt{49 + 36})/2$ and their difference is

   $\dfrac{7 + \sqrt{85}}{2} - \dfrac{7 - \sqrt{85}}{2} = \sqrt{85}$

   or

   $\dfrac{7 - \sqrt{85}}{2} - \dfrac{7 + \sqrt{85}}{2} = -\sqrt{85}.$

   Of the given choices, (E) is correct.

7. $(x^{-1} + y^{-1})^{-1} = \dfrac{1}{(1/x) + (1/y)} = \dfrac{xy}{x + y}.$

8. Two equal circles in a plane cannot have only one common tangent

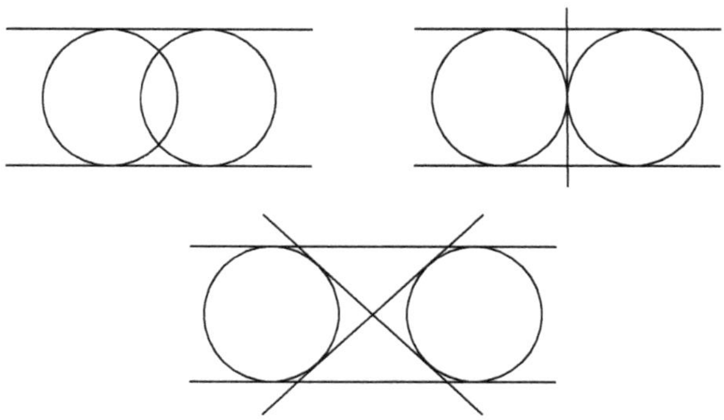

9. $ma - mb = cab$, $ma = b(m + ca)$;

$\therefore \quad b = ma/(m + ca)$.

10. $t_1 = \dfrac{d}{10}$, $t_2 = \dfrac{d}{20}$, $t_1 + t_2 = d\left(\dfrac{1}{10} + \dfrac{1}{20}\right)$;

$$\therefore A = \frac{2d}{d[(1/10) + (1/20)]} = \frac{40}{3} = 13\frac{1}{3} \text{ (mph)};$$

or

$$\frac{1}{A} = \frac{1}{2}\left(\frac{1}{10} + \frac{1}{20}\right); \ \therefore A = \frac{40}{3}.$$

11. The value $x = 1$ makes the denominator zero. Since division by zero is not permitted, $f(1)$ is undefined, so that (C) is the correct choice.

12. Let the first two terms be a and ar where $-1 < r < 1$.

$\therefore a(1 + r) = 4\frac{1}{2}$.

Since $s = a/(1 - r) = 6$, $\quad a = 6(1 - r)$.

$\therefore 6(1 - r)(1 + r) = 4\frac{1}{2}$; $\therefore r = \pm\frac{1}{2}$; $\therefore a = 3$ or 9.

13. The minimum value of the function is at the turning point of the graph where $x = -p/2$. (See 1950, Problem 41.)

14. $12{,}000 = H - \frac{1}{5}H$;

$\therefore H = 15{,}000$, $12{,}000 = S + \frac{1}{5}S$;

$\therefore \quad S = 10{,}000$, $H + S = 25{,}000$.

Therefore, the sale resulted in a loss of \$1000.

15. Since $6^2 + 8^2 = 100 > 9^2$, the triangle is acute, so that (C) is a correct choice. A check of (D) by the Law of Sines or the Law of Cosines shows that it is incorrect. (B) is obviously incorrect by the Law of Sines.

Note : The elimination of (B) and (D) involves trigonometry, and so, in part, this problem is beyond the indicated scope.

Part 2

16. $bh = \left(\frac{11}{10}b\right)(xh)$

$\therefore x = \frac{10}{11}$ and h is reduced by $\frac{1}{11}$ or $9\frac{1}{11}\%$;

or

$bh = \left(b + \frac{1}{10}b\right)\left(h - \frac{r}{100}h\right) = bh + \frac{1}{10}bh - \frac{r}{100}bh - \frac{r}{1000}bh.$

$\therefore \frac{1}{10} - \frac{r}{100} - \frac{r}{1000} = 0$ and $r = 9\frac{1}{11}$ (per cent).

17. Let C be the cost, L the list price, S the selling price and M the marked price.

$C = \frac{4}{5}L, S = C + \frac{1}{5}S, S = \frac{4}{5}M. \therefore \frac{4}{5}\cdot\frac{4}{5}M = \frac{4}{5}L; \therefore M = \frac{5}{4}L;$

$\therefore$(C) is the correct choice.

18. $\log p + \log q = \log(pq)$.

Thus we must have $pq = p + q$.

$\therefore \quad p = q/(q-1).$

19. Since BD bisects angle ABE,

we have $\frac{\overline{AD}}{\overline{DE}} = \frac{\overline{AB}}{\overline{BE}}$;

since BE bisects angle DBC,

we have $\frac{\overline{DE}}{\overline{EC}} = \frac{\overline{BD}}{\overline{BC}}$.

Hence $\frac{\overline{AD}}{\overline{EC}} = \frac{\overline{DE}(\overline{AB}/\overline{BE})}{\overline{DE}(\overline{BC}/\overline{BD})} = \frac{(\overline{AB})(\overline{BD})}{(\overline{BE})(\overline{BC})}$.

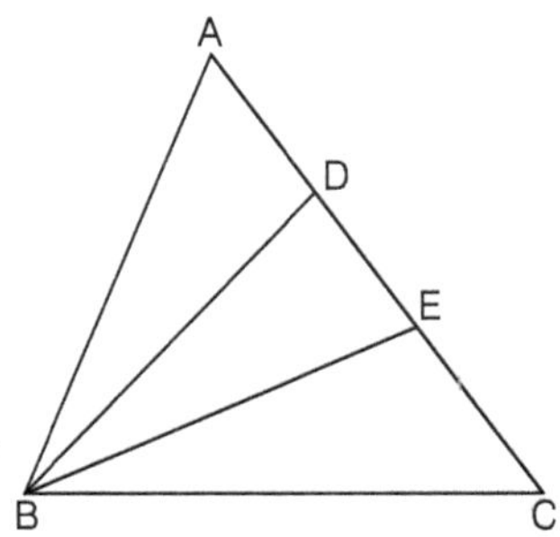

20. $x = \frac{3}{4}y \quad \therefore \frac{x-y}{y} = -\frac{1}{4} \quad \therefore$ (E) is incorrect.

21. Each such angle is the vertex angle of an isosceles triangle whose base angles are each, in degrees, $180 - (n-2)180/n = 360/n$.

$\therefore$angle $= 180 - (720/n) = (180n - 720)/n = 180(n-4)/n.$

22. $x^2 + 4 = b^2 + 1;$

$\therefore \quad x = \sqrt{b^2 - 3}.$

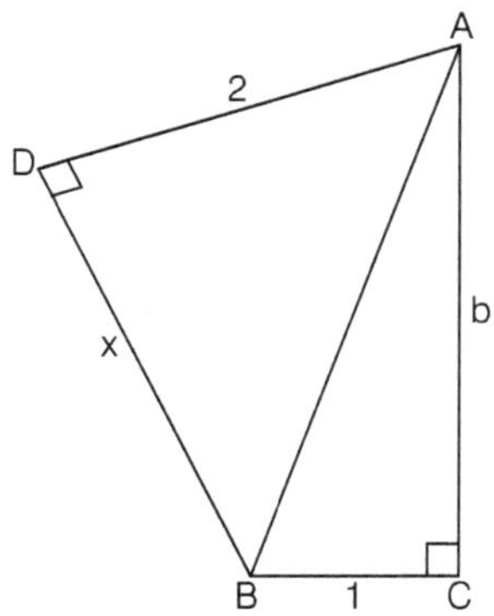

23. The equation is equivalent to $x^2 - \left(b + \frac{m-1}{m+1a}\right)x + c\frac{m-1}{m+1} = 0.$

Since the coefficient of x is equal to the sum of the roots, it must be zero.

We have $b + \frac{m-1}{m+1}a = 0.$

$\therefore \quad bm + b + ma - a = 0;$

$\therefore \quad m = \frac{a-b}{a+b}.$

24. Since $\overline{AB} = 20$ and $\overline{AC} = 12$, $\overline{BC} = 16$. Since $\Delta BDE \sim \Delta BCA$,

$\frac{\text{Area } \Delta BDE}{\text{Area } \Delta BCA} = \frac{10^2}{16^2}$. But Area $\Delta BCA = \frac{1}{2} \cdot 12 \cdot 16 = 96.$

$\therefore$ Area $\Delta BDE = 37\frac{1}{2}$, and the required area is $96 - 37\frac{1}{2} = 58\frac{1}{2}$.

Note : the method here shown is based on subtracting the area of ΔBDE from that of ΔBCA. It is a worthwhile exercise to find the area of the quadrilateral by decomposing it into two right triangles.

25. Since the times are inversely proportional to the rates (see 1951, Problem 39),

$$\frac{t}{30-t} = \frac{1080/3}{8} = 45.$$

$$\therefore \quad t = 30 \cdot \frac{45}{46} \text{ and } d = 8t = 8 \cdot 30 \cdot \frac{45}{46} \sim 245.$$

26. $r^3 + \dfrac{1}{r^3} = \left(r + \dfrac{1}{r}\right)\left(r^2 - 1 + \dfrac{1}{r^2}\right)$.

But $\left(r + \dfrac{1}{r}\right)^2 = r^2 + 2 + \dfrac{1}{r^2} = 3$.

$\therefore r^2 + \dfrac{1}{r^2} = 1;\ \therefore\ r^2 - 1 + \dfrac{1}{r^2} = 0; \therefore r^3 + \dfrac{1}{r^3} = 0.$

27. Let P_1 and P_2 be the perimeters of the smaller and larger triangles respectively. We have

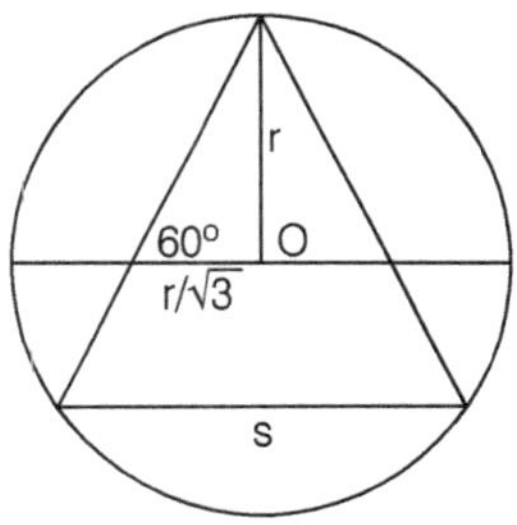

$$r = s\sqrt{3}/2 \text{ and } s = 2r/\sqrt{3}. \therefore\ P_1 = 2r\sqrt{3} \text{ and } P_2 = 3r\sqrt{3};$$

$\therefore P_1 : P_2 = 2:3.$

28. (C) is shown to be the correct choice by direct check.

Note : The fact that the y-differences are 4, 6, 8, and 10 can be used to eliminate (A) and (B) immediately.

29. Let $\overline{KB} = x$ and $\overline{CK} = y$. Then $y(8 - y) = x(10 - x)$ and $y^2 = (5 - x)^2 + 5^2$.

$\therefore 8y - y^2 = 50 - y^2;\ \therefore\ \ y = 25/4.$

$$\therefore \qquad x = \frac{10 - (15/2)}{2} = \frac{5}{4} \text{ and } 10 - x = \frac{35}{4},$$

so that (A) is the correct choice.

30. Using $S = \dfrac{n}{2}[2a + (n-1)d]$, we have $\dfrac{10}{2}(2a + 9d) = 4 \cdot \dfrac{5}{2}(2a + 4d)$.

$\therefore \quad d = 2a;$

$\therefore\ a : d = 1:2.$

31. Single out any one point A. Joined to the remaining 11 points, A yields 11 lines. Since a line is determined by two points, we have for the 12 points, $\dfrac{1}{2}\cdot 12 \cdot 11 = 66$ (lines);

or

$$C(12,2) = \frac{1}{2}\cdot 12 \cdot 11 = 66.$$

32. Time = distance/rate; $\therefore\ t = 30/x$.

33. For a given perimeter the circle has the largest area;

or

let P be the common perimeter, A_1 and A_2 the areas of the circle and the square, respectively. Then

$$P = 2\pi r,\ r = \frac{P}{2\pi},\ A_1 = \frac{P^2}{4\pi};$$

$$P = 4s,\quad s = \frac{P}{4},\quad A_2 = \frac{P^2}{16};$$

Since $4\pi < 16$, $A_1 > A_2$.

34. Let the original price be P.

Then $P_1 = \left(1 + \frac{p}{100}\right)P$.

$$P_2 = \left(1 - \frac{p}{100}\right)P_1 = \left(1 - \frac{p}{100}\right)\left(1 + \frac{p}{100}\right)P = 1;$$

$$\therefore \qquad P = \frac{1}{1 - (p^2/100^2)} = \frac{10{,}000}{10{,}000 - p^2}.$$

35. $$\frac{\sqrt{2}}{(\sqrt{2}+\sqrt{3}) - \sqrt{5}} \cdot \frac{(\sqrt{2}+\sqrt{3}) + \sqrt{5}}{(\sqrt{2}+\sqrt{3}) + \sqrt{5}}$$

$$= \frac{2 + \sqrt{6} + \sqrt{10}}{2\sqrt{6}};\ \frac{2 + \sqrt{6} + \sqrt{10}}{2\sqrt{6}} \cdot \frac{\sqrt{6}}{\sqrt{6}} = \frac{2\sqrt{6} + 6 + 2\sqrt{15}}{2\cdot 6} = \frac{3 + \sqrt{6} + \sqrt{15}}{6}.$$

Part 3

36. $$\frac{x^3+1}{x^2-1} = \frac{(x+1)(x^2-x+1)}{(x+1)(x-1)} = \frac{x^2-x+1}{x-1} \text{ for } x \neq -1.$$

$$\lim_{x\to -1} \frac{x^3+1}{x^2-1} = \lim_{x\to -1} \frac{x^2-x+1}{x-1} = \frac{3}{-2} = -\frac{3}{2}.$$

For continuity at $x = -1$, we must define $\frac{x^3+1}{x^2-1} = -\frac{3}{2}$.

Note : This problem is beyond the indicated scope.

37. By symmetry, the required area is $4(T + S)$.

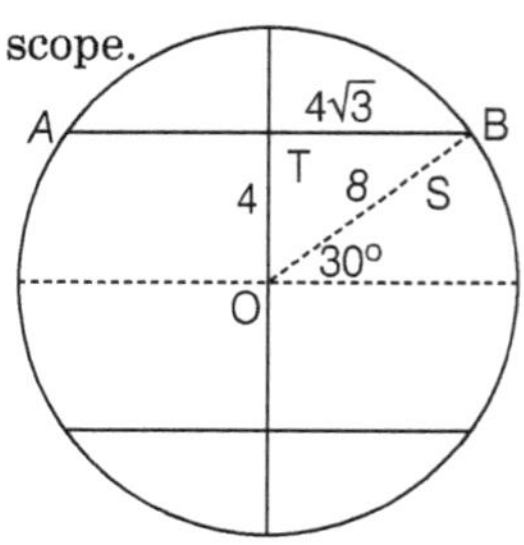

$$T = \frac{1}{2}4\cdot 4\sqrt{3} = 8\sqrt{3},$$

$$S = (30/360)\cdot \pi 8^2 = 5\frac{1}{3}\pi;$$

$$\therefore A = 32\sqrt{3} + 21\frac{1}{3}\pi.$$

38. $1400 = \frac{1}{2} \cdot 50(8a + 8b)$.

$\therefore a + b = 7$. This indeterminate equation is satisfied by the following pairs of integral values : 1, 6; 2, 5; 3, 4.

$\therefore$(D) is the correct choice.

39. $l^2 + w^2 = d^2$, $l + w = p/2$, $(l + w)^2 = l^2 + 2lw + w^2 = p^2/4$,
$2lw = p^2/4 - d^2$, $(l - w)^2 = l^2 - 2lw + w^2 = 2d^2 - p^2/4$,
and $l - w = \sqrt{8d^2 - p^2}/2$.

40. We are told that the values of $f(x)$ listed correspond to

$f(x), f(x + h), f(x + 2h), \dots, f(x + 7h)$.

Observe that the difference between successive values is given by

$$f(x + h) - f(x) = a(x + h)^2 + b(x + h) + c - (ax^2 + bx + c)$$
$$= 2ahx + ah^2 + bh.$$

Since this difference is a linear function of x, it must change by the same amount whenever x is increased by h. But the successive differences of the listed values

	3844	3969	4096	4227	4356	4489	4624	4761
are	125	127	131	129	133	135	137	

so that, if only one value is incorrect, 4227 is that value.

41. $V + y = \pi(r + 6)^2 h = \pi r^2(h + 6)$; $\therefore (r + 6)^2 \cdot 2 = r^2 \cdot 8$

$\therefore \quad 3r^2 - 12r - 36 = 0$; $\therefore r = 6$.

42. $D = .PQQQ\ldots = .a_1\ldots a_r b_1\ldots b_s b_1\ldots b_s\ldots$ So (A), (B), (C) and (E) are all correct choices. To check that (D) is incorrect, we have $10^{r+s}D - 10^r D = PQ - P$. $\therefore 10^r(10^s - 1)D = P(Q - 1)$.

43. For each semi-circle the diameter is $2r/n$, and the length of its arc is $\pi r/n$. The sum of n such arcs is $n \cdot \pi r/n = \pi r =$ semi-circumference.

44. Given $10u + t = k(u + t)$. Let $10t + u = m(u + t)$.

$\therefore \quad 11(t + u) = (k + m)(u + t)$; $\therefore k + m = 11$; $\therefore m = 11 - k$.

45. The Arithmetic Mean is $(a + b)/2$, the Geometric Mean is $\sqrt{ab}$, and the Harmonic Mean is $2ab/(a + b)$. The proper order for decreasing magnitude is (E);

or

Since $(a - b)^2 > 0$, we have $a^2 + b^2 > 2ab$;

$\therefore a^2 + 2ab + b^2 > 4ab$, $a + b > 2\sqrt{ab}$, and $(a + b)/2 > \sqrt{ab}$.

Since $a^2 + 2ab + b^2 > 4ab$, we have

$1 > 4ab/(a + b)^2$;

$\therefore ab > 4a^2b^2/(a + b)^2$, and $\sqrt{ab} > 2ab/(a + b)$.

46. Area (new) $= (d + l)(d - l) = d^2 - l^2 = w^2$;

$\therefore$ (C) is the correct choice.

47. From equation (1) $z = y^2$.

From equation (2) $2^z = 2^{2x+1}$, $z = 2x + 1$; $\therefore\ x = \dfrac{z-1}{2} = \dfrac{y^2-1}{2}$.

From equation (3) $\dfrac{y^2-1}{2} + y + y^2 = 16$. This equation has one integral root $y = 3$.

$\therefore x = 4,\ y = 3,\ z = 9$

48. Let f and s denote the speeds of the faster and slower cyclists, respectively. The distances (measured from the starting point of the faster cyclist) at which they meet are

(1) $f \cdot r = k + s \cdot r$ when they travel in the same direction, and

(2) $f \cdot t = k - s \cdot t$ when the slower one comes to meet the other.

From (2), $f + s = k/t$; from (1), $f - s = k/r$.

$\therefore 2f = \dfrac{k}{r} + \dfrac{k}{t} = \dfrac{k(r+t)}{rt}$, $2s = \dfrac{k}{t} - \dfrac{k}{r} = \dfrac{k(r-t)}{rt}$, and $\dfrac{f}{s} = \dfrac{r+t}{r-t}$.

49. By subtracting from ΔABC the sum of ΔCBF, ΔBAE, and ΔACD and restoring $\Delta CDN_1 + \Delta BFN_3 + \Delta AEN_2$, we have $\Delta N_1N_2N_3$.

$\Delta CBF = \Delta BAE = \Delta ACD = \dfrac{1}{3}\Delta ABC.$

From the assertion made in the statement of the problem, it follows that $\Delta CDN_1 = \Delta BFN_3 = \Delta AEN_2$

$$= \frac{1}{7}\cdot\frac{1}{3}\Delta ABC = \frac{1}{21}\Delta ABC.$$

$$\therefore \Delta N_1 N_2 N_3 = \Delta ABC - 3\cdot\frac{1}{3}\Delta ABC + 3\cdot\frac{1}{21}\Delta ABC = \frac{1}{7}\Delta ABC.$$

50. Rearrange the terms :

$S_1 = 1 + \dfrac{1}{4} + \dfrac{1}{16} + \ldots = \dfrac{4}{3}$ and $S_2 = \dfrac{\sqrt{2}}{4} + \dfrac{\sqrt{2}}{16} + \dfrac{\sqrt{2}}{64} + \ldots = \dfrac{\sqrt{2}}{3}$

$\therefore S = S_1 + S_2 = \dfrac{1}{3}(4 + \sqrt{2}).$

1953 Solutions

Part 1

1. The profit on each orange sold is $4 - 3\frac{1}{3} = \frac{2}{3}$ cent.

$\therefore\ \frac{2}{3}n = 100,\ n = 150.$

2. $D = D_1 + D_2 - D_1D_2,\ D = 20 + 15 - 3 = 32.$

$\therefore S = 250 - (.32)\,(250) = (.68)\,(250),$

that is, 68% of 250. (See 1950, Problem 22.)

3. $x^2 + y^2$ has no linear factors in the real field.

$x^2 + y^2 = (x + iy)\,(x - iy)$ in the complex field.

Note : This problem is probably beyond the indicated scope.

4. Set each factor equal to zero, and solve the three resulting equations. The roots are $0, -4, -4$, and 4. (The factor $x^2 + 8x + 16 = (x + 4)^2$ has the double root -4.)

5. $x = 6^{2.5} = 6^2 \cdot 6^{1/2} = 36\sqrt{6}.$

6. $\frac{5}{2}\,[(5q + 1) - (q + 5)] = \frac{5}{2}\,(4q - 4) = 10(q - 1).$

7. Multiply numerator and denominator by $\sqrt{a^2 + x^2}$ and simplify.

(D) is the correct choice.

8. For intersection $8/(x^2 + 4) = 2 - x.$

$\therefore x^3 - 2x^2 + 4x = x(x^2 - 2x + 4) = 0;\ \therefore\ x = 0.$

9. $\dfrac{4\frac{1}{2} + x}{9 + x} = \dfrac{7}{10};\ \therefore x = 6.$

10. The circumference of the wheel is 6π.

$\therefore N = 5280/6\pi = 880/\pi.$

11. $C_1 = 2\pi r,\ C_2 = 2\pi(r + 10);$

$\therefore C_2 - C_1 = 20\pi \sim 60$ (feet).

12. $A_1/A_2 = \pi 4^2/\pi 6^2 = 4/9.$

13. $A = \frac{1}{2}bh = \frac{1}{2}h(b_1 + b_2);$

$\therefore\ b_1 + b_2 = b = 18.$

$\therefore\quad m = \frac{1}{2}(b_1 + b_2) = 9.$

14. Each assertion is realized in the accompanying figure. Hence, none of these statements is false.

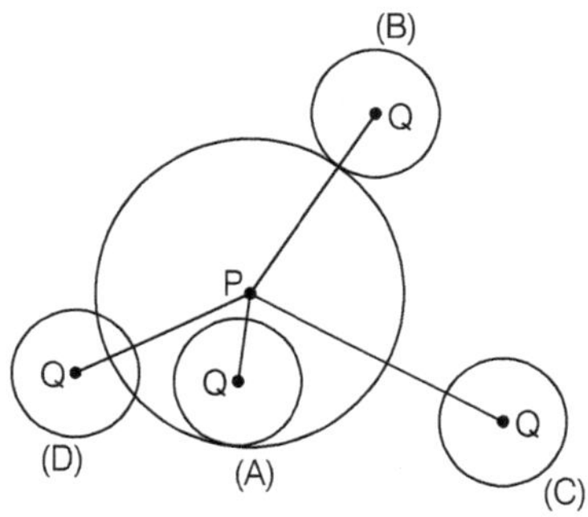

15. Designate a side of the original square by s. Then the radius r of the inscribed circle is $s/2$. A side of the second square (inscribed in the circle) is $r\sqrt{2} = s/\sqrt{2}$. Therefore, the area of the second square is $s^2/2$, while that of the original square is s^2.

$\therefore$(B) is the correct choice.

Part 2

16. S = Selling price = cost + profit + expense $= C + .10S + .15S;$

$\therefore\ S = 4/3C;\quad \therefore$(D) is the correct choice.

17. $\frac{1}{x} = \frac{1}{2}\left(\frac{1}{6} + \frac{1}{4}\right)\ \therefore\ x = 4.8;$

or

Let y be the amount invested at 4%. Then $.04y = .06(4500 - y)$ and $y = 2700$. Therefore the total interest is

$.04(2700) + .06(1800) = 216,$

and $(216/4500)\cdot 100 = 4.8\%.$

18. $x^4 + 4 = (x^4 + 4x^2 + 4) - 4x^2$

$= (x^2 + 2)^2 - (2x)^2$

$= (x^2 + 2x + 2)(x^2 - 2x + 2).$

19. $\frac{3}{4}x\cdot\left(\frac{3}{4}y\right)^2 = \frac{27}{64}xy^2$, and the decrease is $xy^2 - \frac{27}{64}xy^2 = \frac{37}{64}xy^2$.

20. $x^4 + x^3 - 4x^2 + x + 1 = x^2\left[\left(x^2 + \frac{1}{x^2}\right) + \left(x + \frac{1}{x}\right) - 4\right] = 0.$

Since $x^2 + \frac{1}{x^2} = \left(x + \frac{1}{x}\right)^2 - 2$, we have

$x^2(y^2 - 2 + y - 4) = x^2(y^2 + y - 6) = 0.$

21. $x^2 - 3x + 6 = 10^1 = 10$ or $x^2 - 3x - 4 = 0$; $\therefore x = 4$ or -1

22. $27\cdot\sqrt[4]{9}\cdot\sqrt[3]{9} = 3^3\cdot 3^{1/2}\cdot 3^{2/3} = 3^{25/6}$, $\log_3 3^{25/6} = 4\frac{1}{6}$.

23. Multiply both sides of the equation by $\sqrt{x+10}$.

$$x + 10 - 6 = 5\sqrt{x+10},\ x^2 - 17x - 234 = 0,\ x = 26 \text{ or } -9.$$

-9 fails to check, i.e., it is an extraneous root.

24. $100a^2 + 10a(b + c) + bc = 100a^2 + 100a + bc.$

For equality we must have $b + c = 10$.

25. $ar^n = ar^{n+1} + ar^{n+2}$; $\therefore\ r^2 + r = 1$; $\therefore r = (\sqrt{5} - 1)/2$.

26. Let x be the required length. Then $x^2/15^2 = 2/3$.

$\therefore x^2 = 150$; $\therefore x = 5\sqrt{6}$.

27. $S = \pi + \frac{\pi}{4} + \frac{\pi}{16} + \ldots,$

$S = \pi/[1 - (1/4)] = 4\pi/3.$

28. The bisector of an angle of the triangle divides the opposite sides into segments proportional to the other two sides, i.e., $y/c = x/b$. It follows that $x/b = (x + y)/(b + c) = a/(b + c)$.

29. (D) is the correct choice. Consult sections on approximate numbers for the explanation.

Note : For a full understanding of the topic, a knowledge of calculus is needed.

30. B pays A $9000 - \frac{1}{10}\cdot 9000 = \$8100.$

A pays B $8100 + \frac{1}{10}\cdot 8100 = \$8910.$

Thus A has lost $8910 - 8100 = \$810$.

31. 30 ft. per second corresponds to 1 click per second. In miles per hour, $30\cdot 60^2/5280 = 225/11$ mph corresponds to 1 click per second. 1 mph corresponds to 11/225 clicks per second, which is approximately 1 click in 20 seconds. So x miles per hour corresponds approximately to x clicks in 20 seconds.

32. The diagonals of the quadrilateral formed lie along the two perpendicular lines joining the midpoints of the opposite sides of the rectangle. These diagonals are of different lengths, and they are perpendicular bisectors of each other. Therefore, the figure is a rhombus;

or

the sides of the quadrilateral can be proved equal by congruent triangles. Also it can be proved that no interior angle is a right angle.

33. Let x be one of the equal sides of the triangle. $2p = 2x + x\sqrt{2}$;

$$\therefore x = \frac{2p}{2+\sqrt{2}},\ A = \frac{1}{2}x^2 = \frac{1}{2}\cdot\frac{4p^2}{6+4\sqrt{2}}\cdot\frac{6-4\sqrt{2}}{6-4\sqrt{2}} = p^2(3-2\sqrt{2}).$$

34. Let the 12-inch side be AC.

Since $\measuredangle B = 30°$, $AC = 60°$.

$\therefore \quad \overline{AC} = r$;

$\therefore \quad 2r = d = 24$ (inches)

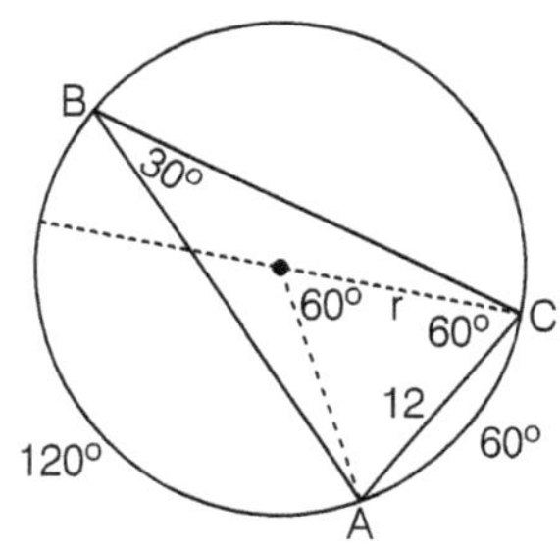

35. $f(x+2) = (x+2)\,(x+2-1)/2$ and $f(x+1) = (x+1)\,(x+1-1)/2$.

So $(x+1)/2 = f(x+1)/x$; $\therefore f(x+2) = (x+2)f(x+1)/x$.

Note : The symbolism here is beyond the indicated scope.

Part 3

36. By actual division, the remainder is found to be $18 + m$. Since $x - 3$ is a factor, $18 + m = 0$, and, therefore, $m = -18$.

$\therefore$(C) is the correct choice,

or

by the Factor Theorem Converse, $36 - 18 + m = 0$. $\therefore$ $m = -18$.

37. Let x be the distance from the center of the circle to the base and let Then

$h = r + x$ be the altitude of ΔABC.

$h^2 + 9 = 144.$

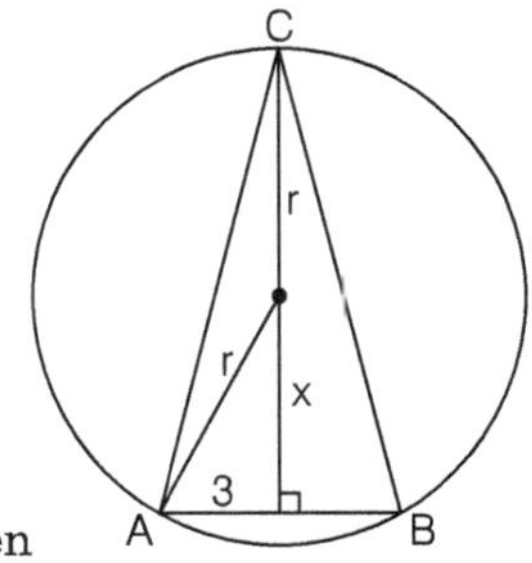

$\therefore \quad h = \sqrt{135}.$

$$r^2 = 3^2 + x^2$$
$$= 3^2 + (h-r)^2$$
$$= 3^2 + h^2 - 2hr + r^2.$$

$$\therefore \quad r = (9 + h^2)/2h$$
$$= (9 + 135)/2\sqrt{135}$$
$$= 144\sqrt{135}/270 = 8\sqrt{15}/5.$$

38. $F[3, f(4)]$ means the value of $b^2 + a$ when $a = 3$ and $b = 2$, namely, 7.

Note : This problem is beyond the indicated scope.

39. For permissible values of a and b, $\log_a b \cdot \log_b a \equiv 1$;

or

Let $x = \log_a b$ and $y = \log_b a$.

$\therefore a^x = b$ and $b^y = a \;\therefore a^{xy} = b^y = a; \quad \therefore\; xy = 1; \;\therefore \log_a b \cdot \log_b a = 1.$

40. To contradict the given statement, preface it with "It is not true that ...". Therefore, the negation is (C);

or

in symbols, given $\forall_x [x \in M \mid x \text{ is honest}]$.

$\therefore \sim \forall_x = \exists_x \; [x \in M \mid x \text{ is dishonest}]$.

41. $\overline{OB} = 400$ rods.

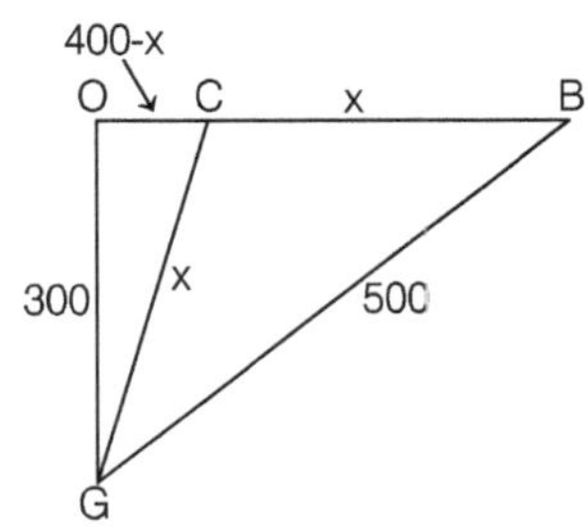

$x^2 = 300^2 + (400 - x)^2,$

$800x = 250{,}000.$

$\therefore \quad x = 312\frac{1}{2}$ (rods);

$\therefore$ (E) is the correct choice.

42. $t^2 = 41^2 - 9^2 = 40^2.$

43. Let s be the price of the article, n the number of sales. Then $[(1 + p)s]\,[(1 - d)n] = sn,\; p - d - pd = 0,\; d = p/(1 + p).$

44. Let the true equation be $x^2 + bx + c = 0$, the equation obtained by the first student be $x^2 + bx + c' = 0$, and the equation obtained by the second student be $x^2 + b'x + c = 0$.

$x^2 + bx + c' = x^2 - 10x + 16 = 0$ and $x^2 + b'x + c = x^2 + 10x + 9 = 0$.

$\therefore x^2 + bx + c = x^2 - 10x + 9 = 0.$

45. For $a \neq b$, Arithmetic Mean > Geometric Mean. For $a = b$, Arithmetic Mean = Geometric Mean. (See 1952, Problem 45.)

$$\therefore \quad \frac{(a+b)}{2} \geq \sqrt{ab}.$$

46. $\sqrt{s^2+l^2} = s + l - \frac{l}{2} = s + \frac{l}{2}, \ \frac{3}{4} l^2 = sl; \quad \therefore \quad \frac{s}{l} = \frac{3}{4}.$

47. For all $x > 0$, $1 + x < 10^x$. $\quad \therefore \log(1 + x) < x$.

Note : This problem may be beyond the indicated scope.

48. $t^2 = s^2 + \left(\frac{t+s}{2}\right)^2,$

$5s^2 + 2ts - 3t^2 = 0$

$= (5s - 3t)(s + t);$

$\therefore \frac{s}{t} = \frac{3}{5}.$

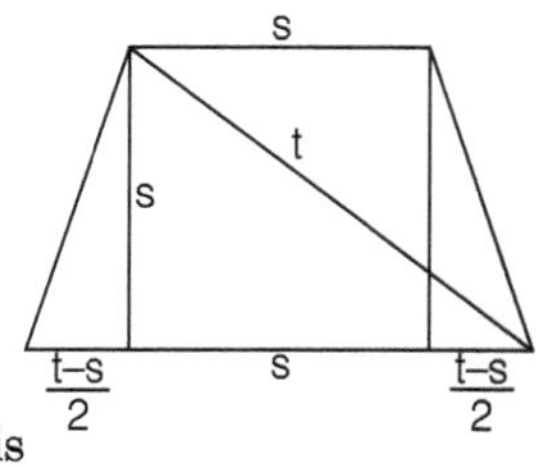

49. The smallest possible value of $\overline{AC} + \overline{BC}$ is obtained when C is the intersection of the y-axis, with the line that leads from A to the mirror image (the mirror being the y-axis) B' : $(-2, 1)$ of B. This is true because $\overline{CB'} = \overline{CB}$ and a straight line is the shortest path between two points. The line through A and B' is given by

$y = \frac{5-1}{5+2}x + k = \frac{4}{7}x + k.$

To find k, we use the fact that the line goes through A:

$5 = \frac{4}{7} \cdot 5 + k, \ k = 5 - \frac{20}{7} = \frac{15}{7} = 2\frac{1}{7} \quad \therefore C$ has coordinates $\left(0, 2\frac{1}{7}\right)$.

50. Denoting the sides of the triangle by a, b, c we observe that $a = 8 + 6 = 14$, $b = 8 + x$, $c = x + 6$.

$\therefore \qquad 2s = a + b + c = 2x + 28, \ s = x + 14.$

On the one hand, the area of the triangle is half the product of the perimeter and the radius of the inscribed circle; on the other hand, it is given in terms of s so that

Area $= r \cdot s = 4(x + 14) = \sqrt{s(s-a)(s-b)(s-c)} = \sqrt{48x(x+14)}$ or $(x+14)^2 = 3x(x+14)$, $x + 14 = 3x$.

$\therefore x = 7$, and the shortest side is $c = 6 + 7 = 13$; or

$\sin\frac{B}{2} = \frac{4}{\sqrt{4^2+6^2}} = \frac{2}{\sqrt{13}}, \ \cos\frac{B}{2} = \frac{3}{\sqrt{13}},$

$\sin B = 2\sin\frac{B}{2}\cos\frac{B}{2} = \frac{12}{13},$

similarly $\sin C = \frac{4}{5}.$

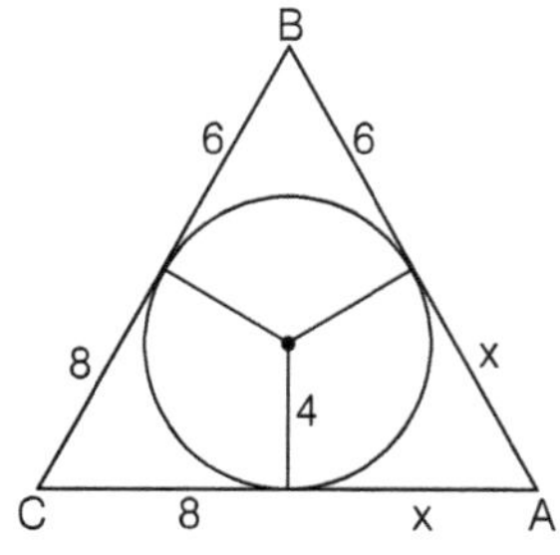

Using the Law of Sines,

$\frac{\sin B}{b} = \frac{\sin C}{c}, \ \frac{12/13}{8+x} = \frac{4/5}{6+x}.$

Thus $x = 7$.

1954 Solutions

Part 1

1. $(5-\sqrt{y^2-25})^2 = 25-10\sqrt{y^2-25}+y^2-25 = y^2-10\sqrt{y^2-25}.$

2. Since $x=1$ is not a solution of the original equation, the solution set of the first equation is the number 4 only;

or

the graph of $y=\dfrac{2x^2}{x-1}-\dfrac{2x+7}{3}+\dfrac{4-6x}{x-1}+1$ is the straight line $y=\dfrac{4}{3}\cdot\dfrac{x^2-5x+4}{x-1}=\dfrac{4}{3}(x-4)$ with the point $(1,-4)$ deleted. Since the graph crosses the x-axis at $(4, 0)$, its only root is 4.

3. $x=k_1y^3$, $y=k_2z^{1/5}$;
$\therefore x=k_1k_2^3z^{3/5}=k_3z^{3/5}$ and (C) is the correct choice.

4. $132=2^2\cdot3\cdot11$ and $6432=2^5\cdot3\cdot67$. Their H.C.D. $=2^2\cdot3=12$ and $12-8=4$.

5. $A=\dfrac{1}{2}ap=\dfrac{1}{2}\cdot5\sqrt{3}\cdot60=150\sqrt{3}$ (square inches).

6. $x^0=1$ for any number $x(x\neq0)$ and $-32=-2^5$.
$\therefore\dfrac{1}{16}+1-\dfrac{1}{\sqrt{64}}-\dfrac{1}{(-2^5)^{4/5}}=\dfrac{1}{16}+1-\dfrac{1}{8}-\dfrac{1}{16}=\dfrac{7}{8}.$

7. Let C be the original price of the dress. Then $25=C-2.50$ or $C=27.50$. $250/2750=1/11\sim9\%$.

8. Let s be the side of the square and h the altitude of the triangle. Then $\dfrac{1}{2}h\cdot2s=s^2$. $\therefore h=s$ or $h/s=1$.

9. $(13+r)(13-r)=16\cdot9$ $\qquad\therefore r^2=25$, $r=5$.

10. Let $a=b=1$.
Then $(1+1)^6=2^6=64$;

or

$1+6+15+20+15+6+1=64.$

11. Let S be the selling price, M the marked price and C the cost. Then $S = M - \frac{1}{3}M = \frac{2}{3}M, C = \frac{3}{4}S = \frac{3}{4}\cdot\frac{2}{3}M = \frac{1}{2}M$ and $\frac{C}{M} = \frac{1}{2}$.

12. The system is inconsistent (see 1950, Problem 14); hence, no solution;

or

graphically, we have two parallel lines;

or

$\begin{vmatrix} 2 & -3 \\ 4 & -6 \end{vmatrix} = 0$; hence no solution.

13. $\measuredangle x = \frac{1}{2}(a + b + c)$.

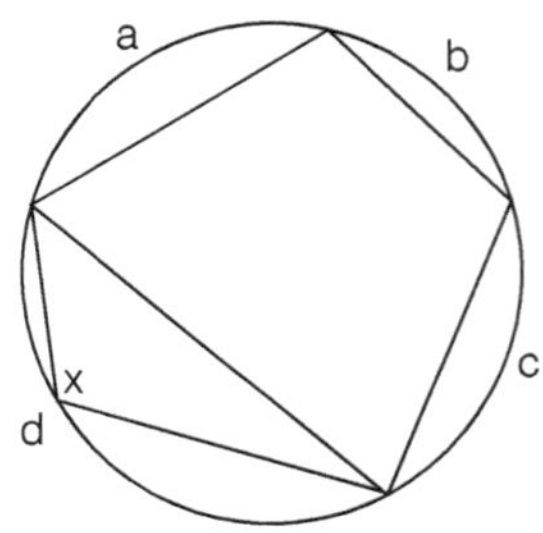

Thus the sum of the four angles

$$S = \frac{1}{2}(a + b + c + b + c + d + c + d + a + d + a + b) = \frac{3}{2}(a + b + c + d) = \frac{3}{2}\cdot 360° = 540°.$$

14. $$\sqrt{1 + \left(\frac{x^4 - 1}{2x^2}\right)^2} = \sqrt{\frac{4x^4 + x^8 - 2x^4 + 1}{4x^4}} = \sqrt{\left(\frac{x^4 + 1}{2x^2}\right)^2} = \frac{x^4 + 1}{2x^2} = \frac{x^2}{2} + \frac{1}{2x^2}.$$

15. $\log 125 = \log(1000/8) = \log 1000 - \log 8 = 3 - \log 2^3 = 3 - 3\log 2$.

Part 2

16. $f(x + h) - f(x) = 5(x + h)^2 - 2(x + h) - 1 - (5x^2 - 2x - 1) = 10xh + 5h^2 - 2h = h(10x + 5h - 2)$.

Note : The symbolism in this problem may be beyond the indicated scope, but the problem itself is quite simple.

17. There are a number of ways of determining that (A) is the correct answer.

One (obvious) way is to plot a sufficient number of points.

Note : Graphs of cubics, generally, are beyond the indicated scope.

18. $2x - 3 > 7 - x$. Adding $x + 3$ to both sides of the inequality, we have $3x > 10$ or $x > \frac{10}{3}$.

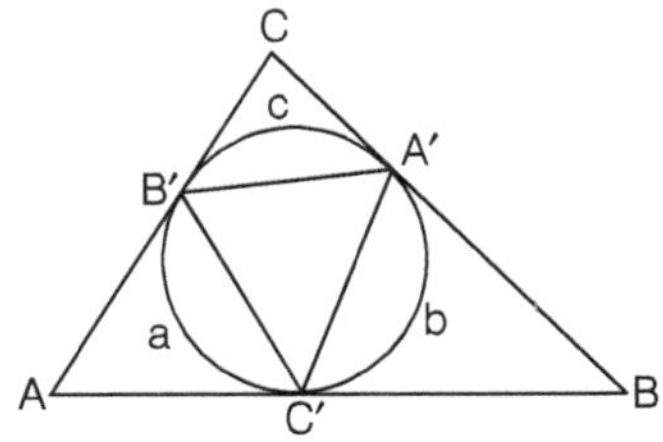

19. $\measuredangle A' = \frac{a}{2}$,

$\measuredangle A = 180° - a$; $\therefore a = 180° - \measuredangle A$.

$\measuredangle A' = \frac{a}{2} = 90° - \frac{\measuredangle A}{2}$;

$\therefore \measuredangle A' < 90°$.

Similarly, $\measuredangle B' < 90°$. $\measuredangle C' < 90°$.

20. By Descartes' Rule of Signs (B) is shown to be correct;

or

solve to find the roots $-1, -2, -3$;

or

graph $y = x^3 + 6x^2 + 11x + 6$.

21. $2\sqrt{x} + \frac{2}{\sqrt{x}} = 5$. Squaring, we obtain $4x + 8 + \frac{4}{x} = 25$. Multiply both sides of the equation by x and obtain $4x^2 - 17x + 4 = 0$.

22. Adding, we have

$$\frac{2x^2 - x - 4 - x}{(x+1)(x-2)} = \frac{2(x-2)(x+1)}{(x+1)(x-2)} = 2$$

for values of x other than 2 or -1.

23. $S = C + \frac{1}{n}C$; $\therefore C = \frac{n}{n+1}S$; $\therefore M = \frac{1}{n}C = \frac{1}{n}\cdot\frac{n}{n+1}S = \frac{S}{n+1}$

24. $2x^2 - x(k-1) + 8 = 0$.

For real, equal roots, the discriminant $(k-1)^2 - 64 = 0$ and $k - 1 = \pm 8$.

$\therefore k = 9$ or -7.

25. The product of the roots is $c(a-b)/a(b-c)$. Since one root is 1, the other is the fraction shown.

26. Let $x = \overline{BD}$ and let r be the radius of the small circle. Draw the line from the center of each of the circles to the point of contact of the tangent and the circle. By similar triangles,

$$\frac{x+r}{r} = \frac{x+5r}{3r}$$

$$\therefore x = r.$$

27. $\frac{1}{3}\pi r^2 h = \frac{1}{2}\cdot\frac{4}{3}\pi r^3$; $\quad \therefore \frac{h}{r} = \frac{2}{1}$.

28. $\frac{mr}{nt} = \frac{4}{3} \cdot \frac{9}{14} = \frac{6}{7}$;

$\therefore mr = 6k$ and $nt = 7k$.

$\therefore \frac{3mr - nt}{4nt - 7mr} = \frac{18k - 7k}{28k - 42k} = \frac{11k}{-14k} = -\frac{11}{14}$.

29. In any right triangle (see accompanying figure), we have $x \cdot c = a^2$ and $(c - x)c = b^2$.

Since $\frac{a}{b} = \frac{1}{2}$, it follows that

$$\frac{x}{c - x} = \frac{x \cdot c}{(c - x) \cdot c}$$

$$= \frac{a^2}{b^2} = \frac{1}{4}.$$

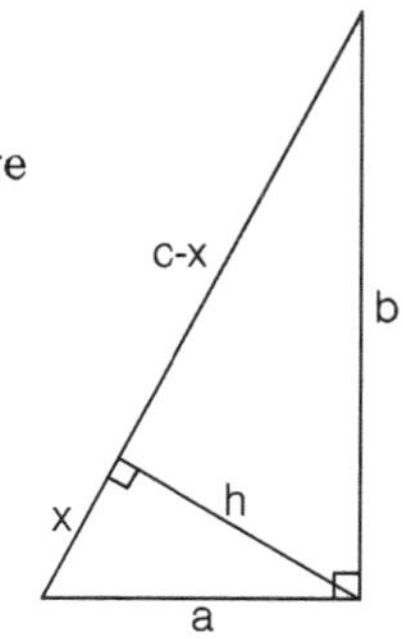

30. Denote by a, b, c the respectively number of days it would take A, B, C to complete the job if he were working alone; then $1/a$, $1/b$, $1/c$ denote the fractions of the job done by each in one day. Since it takes A and B together two days to do the whole job, they do half the job in one day, i.e.,

$\frac{1}{a} + \frac{1}{b} = \frac{1}{2}$, and similarly, $\frac{1}{b} + \frac{1}{c} = \frac{1}{4}$, $\frac{1}{a} + \frac{1}{c} = \frac{5}{12}$.

$\therefore \frac{1}{a} - \frac{1}{b} = \frac{1}{6}$, $\frac{2}{a} = \frac{2}{3}$, $a = 3$.

31. $\measuredangle BOC = 180 - \frac{1}{2}(\measuredangle B + \measuredangle C)$. But $\frac{1}{2}(\measuredangle B + \measuredangle C) = \frac{1}{2} \cdot 140 = 70$.

$\therefore \measuredangle BOC = 110$ (degrees).

32. $x^4 + 64 = (x^4 + 16x^2 + 64) - 16x^2 = (x^2 + 8)^2 - (4x)^2$

$= (x^2 + 4x + 8)(x^2 - 4x + 8)$.

(See also 1953, Problem 18.)

33. The exact interest formula is complex. Approximately, he has the use of \$59 for 1 year. $6 \cdot 100/59 = 10 +$.

$\therefore$ (D) is the correct choice.

34. 1/3 is the infinite repeating decimal 0.333 … . Thus 1/3 is greater than 0.33333333 by the amount $0.00000000333\ldots = 1/3 \cdot 10^8$.

or

$$0.33333333 = 3 \cdot 10^{-1} + 3 \cdot 10^{-2} + \ldots + 3 \cdot 10^{-8}$$

$$= \frac{3 \cdot 10^{-1}(1 - 10^{-8})}{1 - 10^{-1}} = \frac{3(1 - 10^{-8})}{9}$$

$$= \frac{1}{3} - \frac{10^{-8}}{3} = \frac{1}{3} - \frac{1}{3 \cdot 10^8}.$$

35. $x+\sqrt{d^2+(h+x)^2}=h+d,$

$$\sqrt{d^2+(h+x)^2}=h+d-x$$

$d^2+h^2+2hx+x^2, h^2+d^2+x^2+2hd-2hx-2dx,$

$2hx+dx=hd; \therefore x=hd/(2h+d).$

Part 3

36. $\frac{2}{x}=\frac{1}{20}+\frac{1}{10}, x=\frac{40}{3}, \frac{x}{15}=\frac{40/3}{15}=\frac{8}{9}.$

(See also 1951, Problem 39.)

37. $\measuredangle m=\measuredangle p+\measuredangle d,$

$\measuredangle d=\measuredangle q-\measuredangle m$ (there are two vertical angles each $\measuredangle m$).

$\therefore \measuredangle m=\measuredangle p+\measuredangle q-\measuredangle m;$

$\therefore \measuredangle m=\frac{1}{2}(\measuredangle p+\measuredangle q).$

38. $3^x\cdot 3^3=135; \therefore 3^x=5; \ \therefore x\log 3=\log 5=\log(10/2)=\log 10-\log 2.$

$\therefore x=(1-0.3010)/0.4771\sim 1.47.$

39. Let PA be any line segment through P such that A lies on the given circle with center O and radius r. Let A' be the midpoint of PA and let O' be the midpoint of PO. To see that the required locus is a circle with center O' and radius $\frac{1}{2}r$, consider the similar triangles POA and $PO'A'$. For any point A on the given circle, $O'A'=\frac{1}{2}OA=\frac{1}{2}r.$

$\therefore$(E) is the correct choice.

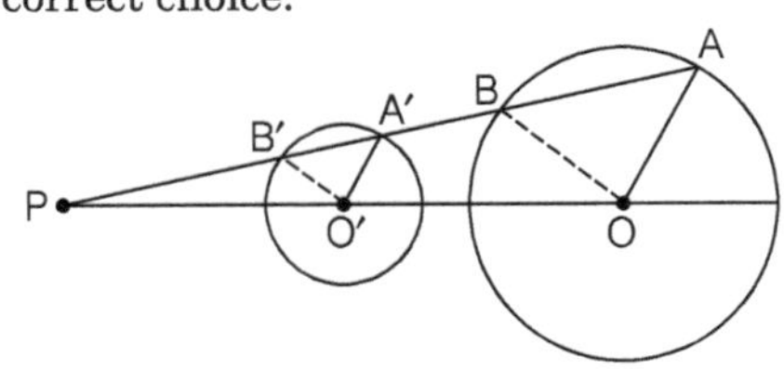

40. $a^3+\frac{1}{a^3}=\left(a+\frac{1}{a}\right)^3-3\left(a+\frac{1}{a}\right)=3\sqrt{3}-3\sqrt{3}=0.$

41. $S=-(-8/4)=2;$

or

find one root, -3, by trial. The reduced equation is $4x^2-20x-3=0$, where the sum of the roots is 5.

$\therefore S=5-3=2.$

42. From the graphs, or from a consideration of the positions of the lowest points, it is found that (D) is the correct answer. (See 1950, Problem 41.)

43. $c = a - r + b - r$ (see 1950, Problem 35). $\therefore 10 = a + b - 2$;

$\therefore P = a + b + c = 10 + 2 + 10 = 22$.

44. We seek an integer whose square is between 1800 and 1850. There is one, and only one, such integer, namely $1849 = 43^2$.

$\therefore 1849 - 43 = 1806$, the year of birth.

or

let y be an integer 0 and 50.

$\therefore 1800 + y + x = x^2$. For x to be integral, the discriminant $1 + 7200 + 4y$ must be an odd square. The value $y = 6$ makes the discriminant $7225 = 85^2$. The year of birth is, therefore, $1800 + 6 = 1806$.

45. Let d denote the distance from A measured along AC and let $l(d)$ denote the length of the segment parallel to BD and d units from A. Then by similar triangles, we have:

For $d \le \dfrac{\overline{AC}}{2}$, $\dfrac{(1/2)l}{d} = \dfrac{(1/2)\overline{BD}}{(1/2)\overline{AC}}$, or $l = 2kd$ where $k = \dfrac{\overline{BD}}{\overline{AC}} =$ constant.

For $d \ge \dfrac{\overline{AC}}{2}$.

$\dfrac{(1/2)l}{AC - d} = \dfrac{(1/2)\overline{BD}}{(1/2)\overline{AC}}$, or $l = 2k(\overline{AC} - d) = -2kd + 2k\overline{AC}$.

The graph of l as a function of d evidently is linear in d. Its slope $2k$ is positive for $d < \overline{AC}/2$; its slope $-2k$ is negative for $d > \overline{AC}/2$. Hence (D) is the correct choice.

46. The distance from the center of the circle to the intersection point of the tangents is 3/8.

$\therefore \overline{CD} = \dfrac{3}{8} + \dfrac{3}{16} = \dfrac{9}{16} = x + \dfrac{1}{2}$;

$\therefore \quad x = \dfrac{1}{16}$ (inch).

47. Since $\overline{MT} = \dfrac{\sqrt{p^2 - 4q^2}}{2}$,

$\overline{AT} = \overline{AM} + \overline{MT} = \dfrac{p + \sqrt{p^2 - 4q^2}}{2}$ and $\overline{TB} = \overline{AB} - \overline{AT}$

$= \dfrac{p - \sqrt{p^2 - 4q^2}}{2}$.

The required equation is, therefore, $x^2 - px + q^2 = 0$.

48. Let x be the distance from the point of the accident to the end of the trip and R the former rate of the train. Then the normal time for the trip, in hours, is

$$\frac{x}{R}+1=\frac{x+R}{R}.$$

Considering the time for each trip, we have

$$1+\frac{1}{2}+\frac{x}{(3/4)R}=\frac{x+R}{R}+3\frac{1}{2} \text{ and } 1+\frac{90}{R}+\frac{1}{2}+\frac{x-90}{(3/4)R}=\frac{x+R}{R}+3.$$

$\therefore x=540$ and $R=60$;

$\therefore$ the trip is $540+60=600$ (miles).

49. $(2a+1)^2-(2b+1)^2$

$$=4a^2+4a+1-4b^2-4b-1$$
$$=4[a(a+1)-b(b+1)].$$

Since the product of two consecutive integers is divisible by 2, the last expression is divisible by 8.

50. $210+x-12x=84; \therefore 11x=126;$

$\therefore x=126/11,\ 126\cdot 60/11\cdot 30 \sim 23$ and $210+x+84=12x$;

$\therefore 11x=294; \therefore x=294/11,\ 294\cdot 60/11\cdot 30 \sim 53.$

$\therefore$ the times are 7:23 and 7:53.

1955 Solutions

Part 1

1. $\frac{3}{8} = .375; \quad \therefore \frac{3}{8} \cdot 10^{-6} = 0.000000375. \quad \therefore$ (D) is the correct choice.

2. $150° - \frac{25}{60} \cdot 30° = 137\frac{1}{2}^{\circ} = 137°30'.$

3. Let A be the arithmetic mean of the original numbers. Then their sum is $10A$ and $A(\text{new}) = (10A + 200)/10 = A + 20.$

$\therefore$ (B) is the correct choice.

4. Multiplying both sides of the equation by $(x-1)(x-2)$, we have $2x - 2 = x - 2 \quad \therefore x = 0.$

5. $y = k/x^2, 16 = k/1, k = 16; \therefore y = 16/x^2, y = 16/8 \cdot 8 = 1/4;$

$y_1/y_2 = x_2^2/x_1^2, 16/y_2 = 8^2/1^2 \ \therefore \ y_2 = 1/4.$

6. Let n be the number bought at each price, and x be the selling price of each.

$2nx = (10n/3) + 4n; \quad \therefore x = 11/3$, that is, 3 for 11 cents.

7. $W_2 = W_1 - W_1/5 = 4W_1/5. \quad \therefore W_1 = 5W_2/4.$

The increase needed is $W_2/4$ or 25% of W_2.

8. $x^2 - 4y^2 = (x+2y)(x-2y) = 0;$

$\therefore \quad x + 2y = 0$ and $x - 2y = 0.$

Each of these equations represents a straight line.

9. This is a right triangle. For any right triangle it can be shown (see 1950, Problem 35) that $a - r + b - r = c.$

$\therefore 2r = a + b - c = 8 + 15 - 17 = 6$ and $r = 3.$

10. The train is moving for $\frac{a}{40}$ hours and is at rest for nm minutes or $\frac{n \cdot m}{60}$ hours.

$\therefore \frac{a}{40} + \frac{nm}{60} = \frac{3a + 2mn}{120}.$

11. The negation is : It is false that no slow learners attend this school. Therefore, some slow learners attend this school.

12. $\sqrt{5x-1}=2-\sqrt{x-1}$, and $5x-1=4-4\sqrt{x-1}+x-1$.

$\therefore\ 4x-4=-4\sqrt{x-1},\ \ x-1=-\sqrt{x-1}$.

Squaring again, we get $x^2-2x+1=x-1$, $x^2-3x+2=0$, $x=1$ or 2.

Since 2 does not satisfy the original equation, we have one root, $x=1$.

13. $\dfrac{(a^{-2}+b^{-2})(a^{-2}-b^{-2})}{a^{-2}-b^{-2}}=a^{-2}+b^{-2}$.

14. The dimensions of R are $1.1s$ and $0.9s$; its area is $0.99s^2$.

$\therefore R/S=.99s^2/s^2=99/100$.

15. Let x be the radius of the larger circle. Then $\pi r^2/\pi x^2=1/3$, $x=r\sqrt{3}$, $r\sqrt{3}-r=r(\sqrt{3}-1)\sim 0.73r$.

Part 2

16. When $a=4$ and $b=-4$, $a+b=0$. Therefore, the expression becomes meaningless for these values.

17. $\log x-5\log 3=\log(x/3^5)$;

$\therefore 10^{-2}=x/3^5$; $\therefore x=2.43$.

18. For real coefficients a zero discriminant implies that the roots are real and equal.

19. The numbers are 7 and -1; the required equation is, therefore,

$$x^2-6x+7=0.$$

20. The expression $\sqrt{25-t^2}+5$ can never equal zero, since it is the sum of 5 and a non-negative number. (By $\sqrt{25-t^2}$ we mean the positive square root).

$\therefore$(A) is the correct choice.

21. Since $\frac{1}{2}hc=A$, $h=2A/c$.

22. Since a single discount D, equal to three successive discounts D_1, D_2, and D_3, is $D=D_1+D_2+D_3-D_1D_2-D_2D_3-D_3D_1+D_1D_2D_3$ (see also 1950, Problem 22), then the choices are:

$$.20+.20+.10-.04-.02-.02+.004=.424 \text{ and}$$
$$.40+.05+.05-.02-.02-.0025+.001=.4585.$$

The saving is $.0345\cdot 10{,}000=345$ (dollars).

23. The counted amount in cents is $25q+10d+5n+c$. The correct amount is $25(q-x)+10(d+x)+5(n+x)+(c-x)$. The difference is $-25x+10x+5x-x=-11x$.

$\therefore\ 11x$ cents should be subtracted.

24. The graph of $y = 4x^2 - 12x - 1$ is a vertical parabola opening upward with the turning point (a minimum) at $(3/2, -10)$. (See 1950, Problem 41);

or

algebraically, the turning point occurs at

$$x = -b/2a = -(-12/8) = 3/2.$$

For this value of x, $y = 4x^2 - 12x - 1 = -10$.

25. $x^4 + 2x^2 + 9 = (x^4 + 6x^2 + 9) - (4x^2)$

$$= (x^2 + 3)^2 - (2x)^2$$
$$= (x^2 - 2x + 3)(x^2 + 2x + 3).$$

$\therefore$ (E) is the correct choice.

26. See 1951, Problem 5.

27. $r^2 + s^2 = (r + s)^2 - 2rs = p^2 - 2q$.

28. For $x \neq 0$, $ax^2 + bx + c \neq ax^2 - bx + c$;

for $x = 0$, $ax^2 + bx + c = ax^2 - bx + c$.

$\therefore$ There is one intersection point $(0, c)$;

$\therefore$ (E) is the correct choice.

29. First, draw the line connecting P and R and denote its other intersections with the circles by M and N; see accompanying figure. The arcs MR and NR contain the same number of degrees; so we may denote each arc by x. To verify this, note that we have two isosceles triangles with a base angle of one equal to a base angle of the other.

$\therefore \measuredangle NOR = \measuredangle MO'R$

$$\measuredangle APR = \frac{1}{2}\{(c + a + c - x) - a\}$$
$$= \frac{1}{2}\{2c - x\}$$
$$\measuredangle BPR = \frac{1}{2}\{b + d + d - (b - x)\}$$
$$= \frac{1}{2}\{2d + x\}$$

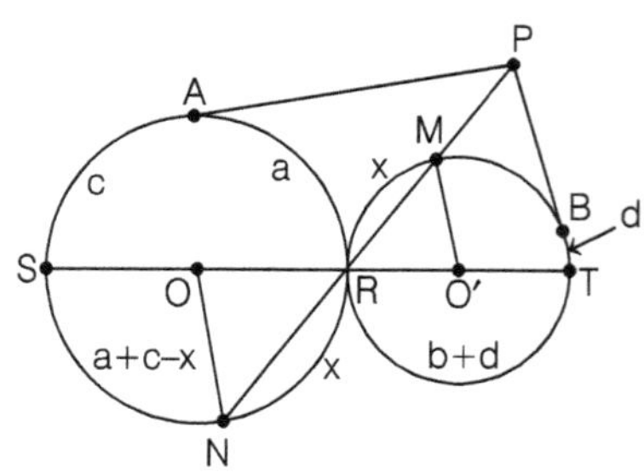

and the sum of angles APR and BPR is $\angle BPA = c + d$.

The desired angle is

$360° - \angle BPA = 360° - (c + d) = (180° - c) + (180° - d) = a + b$.

30. The real roots are, respectively : ± 3; 0 and 2/3; and 3.

$\therefore$ (B) is the correct choice.

31. Let A_1 and A_2 denote the areas of the small and the original triangle, respectively. The median m of the trapezoid is the arithmetic mean of the parallel sides, that is, $m = \frac{1}{2}(b+2)$, where b denotes the shorter parallel side of the trapezoid. To find b, observe that

$$\frac{A_1}{A_2} = \frac{b^2}{2^2} = \frac{1}{2};\ \therefore b = \sqrt{2} \text{ and } m = \frac{1}{2}(\sqrt{2}+2).$$

32. $D = 4b^2 - 4ac = 0; \therefore b^2 - ac = 0; \therefore b^2 = ac,\ a/b = b/c.$

33. Let x be the number of degrees the hour hand moves in the time interval between 8 a.m. and the beginning of the trip, and let y be the number of degrees it moves between 2 p.m. and the end of the trip. The number of degrees traversed by the minute hand during these intervals is $240 + x$ and $60 + y + 180$, respectively. Since the minute hand traverses 12 times as many degrees as the hour hand in any given interval, we have

$$12x = 240 + x,$$

$$\therefore x = \frac{240}{11} \text{ and } 12y = 60 + y + 180,$$

$$\therefore y = \frac{240}{11};$$

that is, the hour hand is just as many degrees past the 8 at the beginning as it is past the 2 at the end of the trip. Since the 8 and the 2 are 180° apart, the initial and final positions of the hour hand differ also by 180°, so that (A) is the correct choice.

34. The shortest length consists of the two external tangents t and the two circular arcs l_1 and l_2.

$$t = 6\sqrt{3},\ 2t = 12\sqrt{3},\ l_1 = \frac{120}{360}\cdot 2\pi\cdot 3 = 2\pi,$$

$$l_2 = \frac{240}{360}\cdot 2\pi\cdot 9 = 12\pi.$$

$$\therefore \text{length} = 12\sqrt{3} + 14\pi.$$

35. The first boy takes $\frac{n}{2} + 1$ marbles, leaving $\frac{n}{2} - 1$ marbles. The second boy takes $\frac{1}{3}\left(\frac{n}{2} - 1\right)$. The third boy, with twice as many, must necessarily have $\frac{2}{3}\left(\frac{n}{2} - 1\right)$, so that n is indeterminate; i.e. n may be any even integer of the form $2 + 6a$, with $a = 0, 1, 2, \ldots$.

Part 3

36. The area of the rectangular surface is $10 \cdot 2x = 40$.

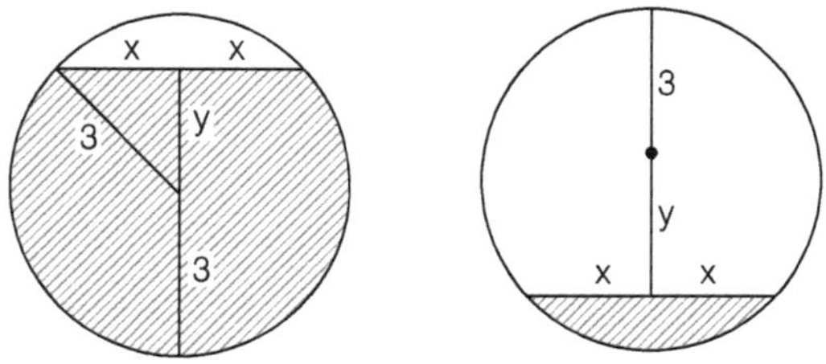

$\therefore x = 2$ and $y = \sqrt{5}$;

$\therefore$ the depth is $3 - \sqrt{5}$, or $3 + \sqrt{5}$.

37. The original number is $100h + 10t + u$. The number with digits reversed is $100u + 10t + h$.

(Since $h > u$, to subtract we must add 10 to u, etc.)

$$\therefore \quad \begin{array}{lll} 100(h-1) & +10(t+9) & +u+10 \\ 100u & +10t & +h \\ \hline 100(h-1-u) & +90 & +10+u-h. \end{array}$$

Since $10 + u - h = 4$, $h - 1 - u = 5$ and therefore, the digits from right to left are 9 and 5.

38. Solving the system;

$$\frac{1}{3}(a+b+c)+d=29$$
$$\frac{1}{3}(b+c+d)+a=23$$
$$\frac{1}{3}(c+d+a)+b=21$$
$$\frac{1}{3}(d+a+b)+c=17,$$

we obtain $a = 12$, $b = 9$, $c = 3$, $d = 21$. Thus (B) is the correct choice.

39. The least value occurs when $x = -p/2$.

For $x = -p/2$, $y = (-p^2/4) + q = 0$;

$\therefore \quad q = p^2/4$ (See 1950, Problem 41).

40. A non-unit fraction b/d is changed in value by the addition of the same non-zero quantity x to the numerator and denominator.

$\therefore$ (A) is the correct choice.

41. $1+\frac{1}{2}+\frac{d-R}{(4/5)R}$

$$=1+1+\frac{80}{R}+\frac{1}{2}+\frac{d-R-80}{(4/5)R};$$

$\therefore \quad R=20.$

(See also 1954, Problem 48.)

42. If $\sqrt{a+\frac{b}{c}}=a\sqrt{\frac{b}{c}}$ then $a+\frac{b}{c}=a^2\frac{b}{c}$.

$\therefore ac=b(a^2-1);$

$\therefore c=b(a^2-1)/a.$

43. $y=(x+1)^2$ and $y(x+1)=1;$

$\therefore (x+1)^2=1/(x+1)$; and

$\therefore (x+1)^3=1$. This last equation has one real and two imaginary roots.

$\therefore$(E) is the correct choice.

44. $\measuredangle OBA=2y;\ \therefore \measuredangle OAB=2y;$ and

$\therefore x=\measuredangle OAB+y=3y.$

45. Let the two series be a, ar, ar^2, ... and $0, d, 2d, \ldots, a+0=1;$

$\therefore a=1$ and $r+d=1$, $r^2+2d=2$.

$\therefore r=2, d=-1.$

$$S_1=\frac{a(r^n-1)}{r-1}=\frac{1(2^{10}-1)}{2-1}=1023,$$

$$S_2=\frac{n}{2}[0+(n-1)d]=\frac{10}{2}[0+9(-1)]=-45.$$

$\therefore S=1023-45=978.$

46. We are required to solve the system of equation :

$$2x+3y=6$$
$$4x-3y=6$$
$$x=2$$
$$y=\frac{2}{3}.$$

Solve the first two simultaneously, obtaining $x=2$, $y=2/3$. Since these results are consistent with the last two equations, the solution is (2, 2/3) and (B) is the correct choice.

47. For the equality $a+bc=a^2+ab+ac+bc$ to hold, we must have $a=a^2+ab+ac$ or $1=a+b+c$.

48. (A) is true because FH is parallel and equal to AE. (C) is true because when HE, which is parallel to CA, is extended, it meets AB in D; DC and BH are corresponding sides of congruent triangles ACD and HDB. (D) is true because

$$\overline{FG} = \overline{FE} + \overline{EG} = \overline{AD} + \frac{1}{2}\overline{DB} = \frac{3}{4}\overline{AB}.$$

(E) is true because G is the midpoint of HB. (FE is parallel to AB and E is the midpoint of CB.) (B) cannot be proved from the given information. Challenge: What additional information is needed to prove (B)?

49. Since $y = (x^2-4)/(x-2) = (x-2)(x+2)/(x-2) = x+2$ (for $x \neq 2$, i.e. $y \neq 4$), then $y = (x^2-4)/(x-2)$ is a straight line with the point (2, 4) deleted. The straight line $y = 2x$ crosses the first line in the deleted point.

∴ (C) is the correct choice.

50. Let r be the increase in the rate of A and d be the distance (in miles) A travels to pass B safely. The times traveled by A, B and C are equal :

$$\frac{d}{50+r} = \frac{d-(30/5280)}{40},$$

$$\frac{d-(30/5280)}{40} = \frac{(210/5280)-d}{50},$$

Solve the second equation for d.

$\therefore d = 110/5280$ miles.

Solve the first equation for r.

$\therefore r = 5$ mph.

1956 Solutions

Part 1

1. $2 + 2(2^2) = 2 + 8 = 10.$

2. $S_1 = C_1 + \frac{1}{5}C_1$ and $S_2 = C_2 - \frac{1}{5}C_2$.

$\therefore C_1 = \frac{5}{6}S_1 = 1.00$ and $C_2 = \frac{5}{4}S_2 = 1.50.$

$\therefore S_1 + S_2 = 2.40$ and $C_1 + C_2 = 2.50;$

$\therefore$ (D) is the correct choice.

3. $5{,}870{,}000{,}000{,}000 = 587 \cdot 10^{10}$, $(587 \cdot 10^{10})\,100 = 587 \cdot 10^{12}$.

4. $\frac{1}{20} \cdot 4000 + \frac{1}{25} \cdot 3500 + \frac{x}{100} \cdot 2500 = 500,\ 25x = 160,\ x = 6.4.$

5. Around a given circle can be placed exactly six circles each equal to the given circle, tangent to it, and tangent to two others. The arc between two successive points of contact of the given circle and the outside circles is one-sixth of the circumference.

6. Let a be the number of cows and b the number of chickens. Then $4a + 2b$ is the total number of legs and

$4a + 2b = 14 + 2(a + b),\ 2a = 14,\ a = 7$ (cows).

Note : The number of chickens is indeterminate.

7. For reciprocal roots, the product of the roots $c/a = 1$. $\therefore c = a$.

8. $8 \cdot 2^x = 5^0;\ \therefore\ 2^{3+x} = 5^0 = 1;\ \therefore 3 + x = 0$ and $x = -3$.

Note : $2^0 = 5^0$, but, otherwise, $2^a \neq 5^a$.

9. $a^{9(1/6)(1/3)4} \cdot a^{9(1/3)(1/6)4} = a^2 \cdot a^2 = a^4.$

10. $\measuredangle ADB = \frac{1}{2}\measuredangle ACB = 30°.$

11. $\frac{1 \cdot (1 + \sqrt{3})\,(1 - \sqrt{3}) - 1 \cdot (1 - \sqrt{3}) + 1 \cdot (1 + \sqrt{3})}{(1 + \sqrt{3})\,(1 - \sqrt{3})} = 1 - \sqrt{3}.$

12. $x^{-1} - 1 = \frac{1}{x} - 1 = \frac{1 - x}{x},\ \frac{1 - x}{x} \cdot \frac{1}{x - 1} = -\frac{1}{x}.$

13. $y - x$ is the excess of y over x. the basis of comparison is the ratio of the excess to y, namely, $(y-x)/y$.
Therefore, the per cent required is $100(y-x)/y$.

14. $\overline{PB}\cdot\overline{PC} = \overline{PA}^2$. $\overline{PB}(\overline{PB}+20) = 300$;
$\therefore \overline{PB} = 10$.

15. Multiply both sides of the equation by $x^2 - 4 = (x+2)(x-2)$.
$15 - 2(x+2) = x^2 - 4$; $\therefore x^2 + 2x - 15 = 0$;
and $\therefore x = -5$ or $x = 3$. Each of these roots satisfies the original equation.

Part 2

16. Let $10m$, $15m$, $24m$ be the three numbers. Then
$10m + 15m + 24m = 98$,
$\therefore m = 2$ and $15m = 30$.

17. $\dfrac{A(2x-3)+B(x+2)}{(x+2)(2x-3)} = \dfrac{5x-11}{2x^2+x-6}$;
$A(2x-3) + B(x+2) \equiv 5x - 11$.
Equating the coefficients of like powers of x, we obtain $2A + B = 5$, $-3A + 2B = -11$; $\therefore A = 3, B = -1$.

18. $10^{2v} = 5^2$,
$10^v = 5, 10^{-v} = 1/10^v = 1/5$.

19. Let the height in each case be 1.
$1 - \frac{1}{4}t = 2\left(1 - \frac{1}{3}t\right); \therefore t = 2\frac{2}{5}$.

20. $\left(\frac{2}{10}\right)^x = 2; \therefore \log\left(\frac{2}{10}\right)^x = x\log\frac{2}{10}$

$$= x(\log 2 - \log 10) = \log 2.$$

$$\therefore \qquad x = \frac{0.3010}{0.3010 - 1} \sim -0.4.$$

21. Each line has 1 or 2 intersection points. Therefore, for both lines, there may be 2, 3, or 4 intersection points.

22. $t_1 = 50/r$, $t_2 = 300/3r = 100/r = 2t_1$;
$\therefore$ (B) is the correct choice.

23. Each of the roots is $\sqrt{2}/a$.
Since a is real, (C) is necessarily the correct choice.

24. Let $\measuredangle DAE = a$.

Then $\measuredangle BCA = \frac{1}{2}(180 - 30 - a)$

$= 75 - (a/2)$ and $\measuredangle DEA = \frac{1}{2}(180 - a) = 90 - (a/2) = \measuredangle ADE$.

$a + \measuredangle ADE + x + \measuredangle BCA = 180$;

$\therefore x = 15$ (degrees).

25. $s = 3 + 5 + 7 + \ldots + (2n + 1) = \frac{n}{2}(3 + 2n + 1) = n(n + 2)$.

26. Combination (A) determines the shape of the triangle, but not its size. All the other combinations determine both the size and the shape of the triangle.

27. The two triangles are similar and hence $A(\text{new})/A(\text{old}) = (2s)^2/s^2 = 4$.

$\therefore$(C) is the correct choice.

28. Let $4x$ be the daughter's share, $3x$ the son's share. Then

$4x + 3x = \frac{1}{2}$ estate and $6x + 500 = \frac{1}{2}$ estate.

$\therefore 7x = 6x + 500$; $\therefore x = 500$; and $\therefore 13x + 500 = 7000$ (dollars).

29. $x^2 + y^2 = 25$ is a circle with center at the origin. $xy = 12$ is a hyperbola, symmetric with respect to the line $x = y$.

$\therefore$points of intersection are symmetric with respect to $x = y$. There are either no intersections, two intersections (if the hyperbola is tangent to the circle) or four intersections. Simultaneous solution of the equations reveals that there are four points of intersection, and that they determine a rectangle.

30. $h = \frac{s}{2}\sqrt{3}$; $\therefore s = \frac{2h}{\sqrt{3}}$, $A = \frac{s^2\sqrt{3}}{4} = \frac{4h^2}{3}\cdot\frac{\sqrt{3}}{4} = \frac{6\sqrt{3}}{3} = 2\sqrt{3}$.

31. $20 = 1\cdot 4^2 + 1\cdot 4^1 + 0\cdot 4^0$;

$\therefore$the required number is 110_4;

or

$20 \div 4$ yields 5 and remainder 0;

$5 \div 4$ yields 1 and remainder 1;

$\therefore 20_{10}$ becomes 110_4.

32. After $1\frac{1}{2}$ minutes each was in the center of the pool. After 3 minutes they were at opposite ends of the pool. After $4\frac{1}{2}$ minutes each was again in the center of the pool.

33. It can be shown † that $\sqrt{2}$ is not a rational number. It can also be shown† that repeating non-terminating decimals represent rational numbers, and terminating decimals clearly represent rational numbers with a power of 10 in the denominator. Hence a non-terminating, non-repeating decimal is the only possibility for representing $\sqrt{2}$.

34. By considering the special case $n = 2$ we immediately rule out choices (B), (C), (D), and (E). We now show that (A) holds.
$n^2(n^2-1) = n\{(n-1)n(n+1)\} = nk$, where k is a product of three consecutive integers, hence always divisible by 3. If n is even, k is divisible also by 2 (since n is a factor of k), hence by 6, and nk by 12; if n is odd, k is divisible also by 4 (since the even numbers $n-1$ and $n+1$ are factors of k), hence by 12.

35. $A = 16\cdot 8\sqrt{3} = 128\sqrt{3}$.

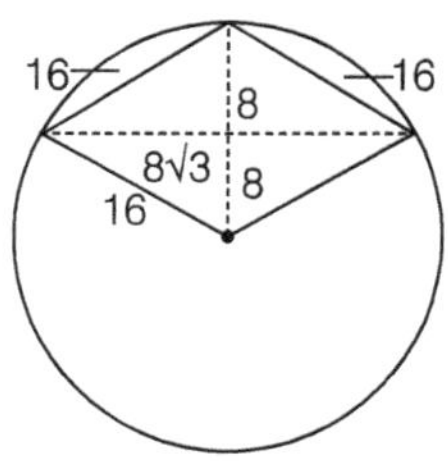

Part 3

36. $S = K(K+1)/2 = N^2$. The possible values for N^2 are $1^2, 2^2, 3^2, \dots, 99^2$. For K to be integral, the discriminant $1+8N^2$ of the equation $K^2+K-2N^2=0$ must be a perfect square. This fact reduces the possible values for N^2 to $1^2, 6^2$, and 35^2. Hence the values of K are 1, 8 and 49. **Note** : There are ways of shortening the number of trials for N^2 still further, but these involve a knowledge of number-theoretic theorems. The shortest way to do this problem is by testing the choices given.

37. The diagonal divides the rhombus into two equilateral triangles.
$A(\text{rhombus}) = 2\cdot s^2\cdot\frac{\sqrt{3}}{4} = 2\cdot\frac{(3/16)^2\sqrt{3}}{4}$. Since the map scale is 400 miles to $1\frac{1}{2}$ inches, 1 inch represents $\frac{2}{3}\cdot 400$ miles and 1 square inch represent $\left(\frac{2}{3}\cdot 400\right)^2$ square miles. The area of the estate is thus $\frac{2\cdot 3^2\sqrt{3}}{16^2\cdot 4}\cdot\frac{800^2}{3^2} = 1250\sqrt{3}$ (square miles).

† See, *e.g.*, Ivan Niven, *Numbers' Rational* and *Irrational*, also in this series.

38. The segments of the hypotenuse are $\frac{a^2}{c}$ and $\frac{b^2}{c}$ (see 1954, Problem 29).

Using similar triangles, $\frac{x}{a}=\frac{b^2/c}{b}, \frac{x}{b}=\frac{a^2/c}{a}$.

$\therefore x^2=\frac{a^2}{c}\cdot\frac{b^2}{c}=\frac{a^2b^2}{a^2+b^2}$; and $\therefore \frac{1}{x^2}=\frac{1}{a^2}+\frac{1}{b^2}$.

39. $b^2=c^2-a^2=c^2-(c-1)^2=2c-1=c+c-1=c+a;$

or

$b^2=c^2-a^2=(a+1)^2-a^2=2a+1=a+1+a=c+a;$

or

$b^2=c^2-a^2=(c+a)(c-a)=c+a$ since $c-a=1$.

40. $2S=gt^2+2V_0t;\ Vt=gt^2+V_0t;\ \therefore 2S-Vt=V_0t;$ and

$\therefore t=\frac{2S}{V+V_0};$

or

$g=\frac{V-V_0}{t}, S=\frac{1}{2}\frac{V-V_0}{t}\cdot t^2+V_0t=t\left(\frac{1}{2}V+\frac{1}{2}V_0\right).$

$\therefore t=\frac{2S}{V+V_0}.$

41. Since $y=2x$, $3y^2+y+4=12x^2+2x+4$.

$\therefore 12x^2+2x+4=12x^2+4x+4;$

$\therefore \quad 2x=4x;$

$\therefore \quad x=0.$

42. $\sqrt{x+4}=\sqrt{x-3}-1;$

$\therefore x+4=x-3-2\sqrt{x-3}+1;$

$\therefore 3=-\sqrt{x-3}.$

This is impossible since the left side is positive while the right side is negative.

43. Let the largest side be c. Then $a+b+c\le 12$.

But $c<a+b$.

$\therefore 2c<12$ or $c<6$. Now try integral combinations such that the triangle is scalene. Since c is the largest side, it cannot be less than 4.

$\therefore$ there are 3 combinations:

$c=5\ \ c=5\ c=4,\ a=4\ a=4\ a=3,\ b=3\ b=2\ b=2$

44. Since $x<a$ and $a<0$, $x^2>ax$ and $ax>a^2$.

$\therefore x^2>ax>a^2.$

45. $N_1 = \dfrac{1 \text{ mile}}{\pi D}$, $N_2 = \dfrac{1\text{mile}}{\pi[D-(1/2)]}$,

$$N_2 - N_1 = \frac{1 \text{ mile}}{\pi}\cdot\frac{1/2}{D[D-(1/2)]}. \therefore \frac{N_2 - N_1}{N_1} = \frac{1/2}{D-1/2} = \frac{1/2}{24\frac{1}{2}} = \frac{1}{49};$$

$\therefore$ the percent increase is $100\cdot\dfrac{1}{49} \sim 2\%$.

46. Solving for x we have $x = 1/(2N+1)$. Since N is positive.
$1/(2N+1)$ is positive and less than 1.
$\therefore$ (A) is the correct choice.

47. In 1 day, x machines can do $\dfrac{1}{3}$ of the job and $x+3$ machines $\dfrac{1}{2}$ of the job.
$\therefore$ 3 machines can do $\dfrac{1}{2} - \dfrac{1}{3} = \dfrac{1}{6}$ of the job in 1 day, and 1 machine can do $\dfrac{1}{18}$ of the job in 1 day. hence, 18 days are needed for 1 machine to complete the job;

or

$(x+3)/x = 3/2$;
$\therefore x = 6$, that is, 6 machines can do 1/3 of the job in 1 day, etc.

48. Let $(3p+25)/(2p-5) = n$, a positive integer. Then, $3p + 25 = kn$, $2p - 5 = k$;
$\therefore k(2n-3) = 65 = 1\cdot 65 = 5\cdot 13$.
$\therefore k = 1, 65, 5$, or 13 and $2n - 3 = 65, 1, 13$, or 5 correspondingly.
$\therefore 2p = 5+1, 5+65, 5+5$, or $5+13$;
$\therefore p = 3, 35, 5$, or 9.
$\therefore$ choice (B) is correct.

49. $\measuredangle P = 40°$;
$\therefore \measuredangle PAB + \measuredangle PBA = 180° - 40° = 140°$.
$\measuredangle TAS = 180° - \measuredangle PAB$;

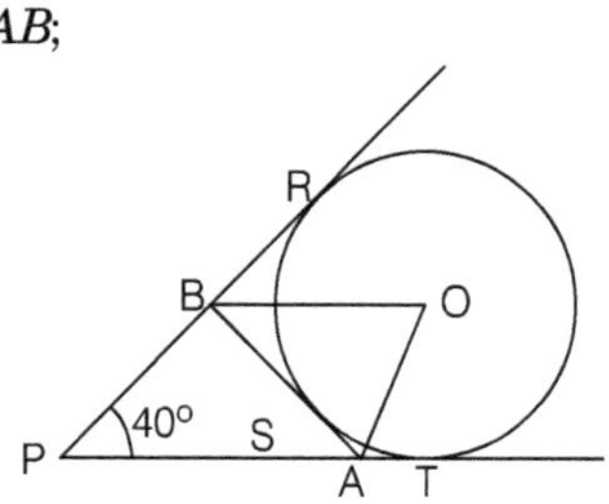

$\measuredangle RBS = 180° - \measuredangle PBA;$

$\measuredangle TAS + \measuredangle RBS = 360° - 140° = 220°.$

Since OA and OB bisect angles TAS and RBS, respectively,

$$\measuredangle OAS + \measuredangle OBS = \frac{1}{2}(220°) = 110°.$$

$$\therefore \measuredangle AOB = 180° - 110° = 70°.$$

The number of degrees in $\measuredangle AOB$ is independent of the position of tangent ASB.

50. $\measuredangle ACB = 180° - 2A.$

Since $\overline{CA} = \overline{CD}$,

$$\measuredangle CAD = \frac{1}{2}[180° - (\measuredangle ACB + 90°)]$$

$$= \frac{1}{2}[180° - (180° - 2A + 90°)]$$

$$= A - 45°$$

$$= A - X$$

$$\therefore \quad X = 45°.$$

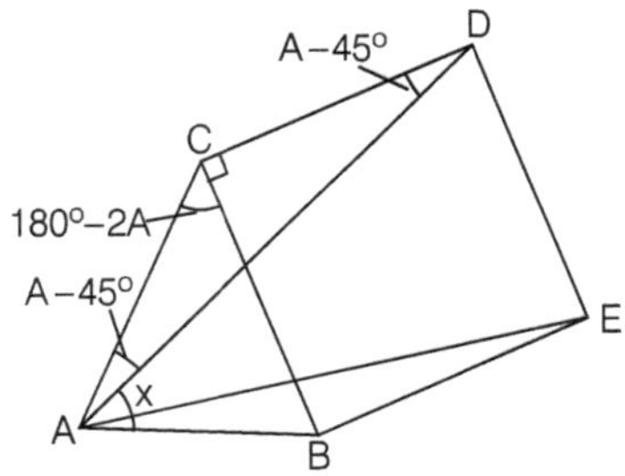

1957 Solutions

Part 1

1. The altitudes, the medians, and the angle –bisectors are distinct for the legs but coincident for the base. Hence, 7 lines.
2. $h/2 = 4$, $h = 8$, $2k/2 = -3$, $k = -3$.
3. $1 - (1 - a)/(1 - a + a) = 1 - 1 + a = a$, provided $a \neq 1$.
4. In this case $a = 3x + 2$, $b = x\,c = -5$.

   $\therefore\ ab + ac = (3x + 2)x + (3x + 2)(-5)$.
5. $\log a - \log b + \log b - \log c + \log c - \log d$

   $$- \log a - \log y + \log d + \log x = \log(x/y).$$
6. The dimensions are x, $10 - 2x$, $14 - 2x$. $\therefore V = 140x - 48x^2 + 4x^3$.
7. For the inscribed circle.

   $r = h/3 = s\sqrt{3}/6$. $p = 3s = 6\sqrt{3}r = (6\sqrt{3})(4\sqrt{3}) = 72$

   $(A = \pi r^2,\ r = \sqrt{48} = 4\sqrt{3})$.
8. The numbers are easily found to be 20, 30 and 50.

   $\therefore 30 = 20a - 10$; $\therefore a = 2$
9. $2 - (-2)^4 = 2 - 16 = -14$.
10. For a low or high point $x = -4/4 = -1$ (see 1950, Problem 41). When $x = -1$, $y = 1$. The point $(-1, 1)$ is a low point since the coefficient of x^2 is positive.
11. $90 - \left(60° + \frac{1}{4}\cdot 30\right)^{\circ} = 22\frac{1}{2}^{\circ}$.

    $\therefore$ (E) is correct. (See also 1951, Problem 12)
12. $10^{-49} = 10\cdot 10^{-50}$, $\therefore 10^{-49} - 2\cdot 10^{-50}$

    $$= 10^{-50}(10 - 2) = 8\cdot 10^{-50}.$$
13. (A) and (B) are irrational. Of the remaining choices only (C) is between $\sqrt{2}$ and $\sqrt{3}$.
14. Since $\sqrt{x^2 - 2x + 1} = |x - 1|$ and $\sqrt{x^2 + 2x + 1} = |x + 1|$, (D) is correct.
15. The formula satisfying the table is $s = 10t^2$. When $t = 2.5$, $s = 62.5$.

Part 2

16. Since the number of goldfish n is a positive integer from 1 to 12, the graph is a finite set of distinct points.

17. The longest path covers 8 edges, and $8 \cdot 3 = 24$ (inches).

18. ABM is a right triangle. By similar triangles, $\overline{AP}/\overline{AB} = \overline{AO}/\overline{AM}$.
$\therefore \overline{AP} \cdot \overline{AM} = \overline{AO} \cdot \overline{AB}$.

19. $10011_2 = 1 \cdot 2^4 + 0 \cdot 2^3 + 0 \cdot 2^2 + 1 \cdot 2^1 + 1 \cdot 2^0 = 16 + 2 + 1 = 19$.
(See also 1956, Problem 31.)

20. $\dfrac{1}{A} = \dfrac{1}{2}\left(\dfrac{1}{50} + \dfrac{1}{45}\right)$; $\therefore A = 47\dfrac{7}{19}$. (See 1950, Problem 27.)

21. (1) is the inverse, (2) is the converse, (3) is the contrapositive, and (4) is an alternative way of stating the theorem.
$\therefore$ the combination (3), (4) is correct.

22. $\sqrt{x-1} + 1 = \sqrt{x+1}$; $\therefore x - 1 + 2\sqrt{x-1} + 1 = x + 1$; $\therefore\ 2\sqrt{x-1} = 1$;
$\therefore\ 4x - 4 = 1$; and $\therefore\ 4x = 5$. (This value of x satisfies the original equation).

23. The intersection points are (0, 10) and (1, 9). $d = \sqrt{(-1)^2 + 1^2} = \sqrt{2}$.

24. $(10t + u)^2 - (10u + t)^2 = 99(t^2 - u^2)$. This is divisible by 9, 11, $t + u$, $t - u$. $\therefore$(B) is the correct choice.

25. From $R : (c, d)$, draw line segment RA perpendicular to the x-axis.
Let O denote the origin (0, 0).

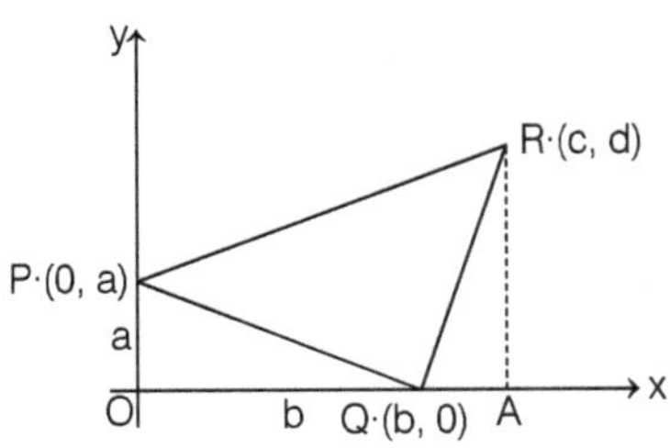

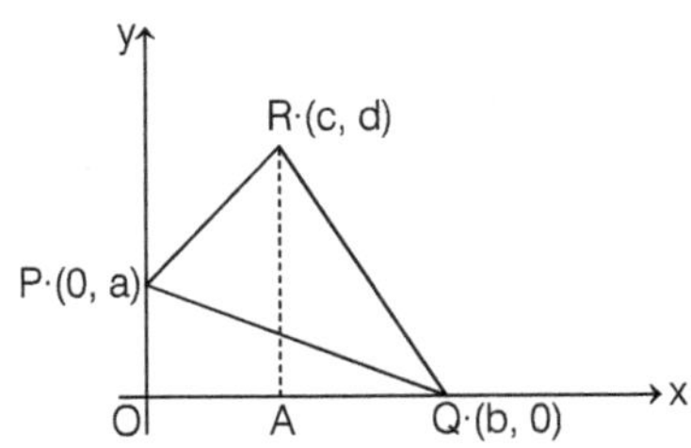

If $c > b$,
area ΔPQR = area trapezoid $OPRA$ − area ΔQAR − area ΔOPQ
$= \dfrac{1}{2}c(a + d) - \dfrac{1}{2}d(c - b) - \dfrac{1}{2}ab = \dfrac{1}{2}(ac + bd - ab)$.

If $c < b$,
area ΔPQR = area trapezoid $OPRA$ + area ΔQAR − area ΔOPQ
$= \dfrac{1}{2}c(a + d) - \dfrac{1}{2}d(b - c) - \dfrac{1}{2}ab = \dfrac{1}{2}(ac + bd - ab)$.

26. Let d be the distance from the point P to any side, say c, of the triangle. Then we want Area $\Delta APB = \frac{1}{3}$ Area ΔABC, or if h is the altitude to the side c, $\frac{1}{2}dc = \frac{1}{3}\cdot\frac{1}{2}ch$. $\therefore d = \frac{1}{3}h$.

For this to hold for each side of the triangle, the required point must be the intersection of the medians.

27. $\frac{1}{r}+\frac{1}{s}=\frac{r+s}{rs}=\frac{-p}{q}$

28. Let $x = b^{\log_b a}$; $\therefore \log_b x = \log_b a \cdot \log_b b = \log_b a$; $\therefore x = a$; and $\therefore$ (B) is correct.

29. First note that the equality $x^2(x^2-1)=0$ is satisfied by $0, 0, +1, -1$.
Since x^2 is non-negative, then $x^2(x^2-1)>0$ implies $x^2-1>0$.
$\therefore x^2>1$; $\therefore |x|>1$, that is, $x>1$ or $x<-1$.
Combining these, we have $x=0$, $x\le -1$, $x\ge 1$.

30. The formula is $S = n(n+1)(2n+1)/6$;

or

first let $S = 1^2$ and then let $S = 1^2+2^2$

$1^2 = \frac{1(1+c)(2+k)}{6}$ and $1^2+2^2 = \frac{2(2+c)(4+k)}{6}$

Solving for c and k, we have $c = k = 1$.

31. Let x be the length of one of the legs of the isosceles triangle. Then a side of the square is given by $1-2x$ and this side must equal the hypotenuse of the triangle (to form a regular octagon).
$\therefore x\sqrt{2} = 1-2x \quad \therefore x = (2-\sqrt{2})/2$.

32. Since $n^5 - n = (n-1)n(n+1)(n^2+1) = k(n^2+1)$, where k is a product of 3 consecutive integers, k is always divisible by 6 at least (see 1951, Problem 15). We show next that in addition, $n^5 - n$ is always divisible by 5. If we divide n by 5, we get a remainder whose value is either 0, 1, 2, 3, or 4; i.e., $n = 5q + r$, $r = 0, 1, 2, 3$ or 4.

By the binomial theorem,

$$n^5 - n = (5q+r)^5 - (5q+r) = (5q)^5 + 5(5q)^4r + 10(5q)^3r^2 + 10(5q)^2r^3 + 5(5q)r^4 + r^5 - 5q - r$$

$= 5N + r^5 - r$, where N is an integer.

If $r=0$,	$n^5 - n = 5N$	is divisible by 5.
If $r=1$,	$n^5 - n = 5N$	is divisible by 5.
If $r=2$,	$n^5 - n = 5N + 30 = 5(N+6)$	is divisible by 5.
If $r=3$,	$n^5 - n = 5N + 240 = 5(N+48)$	is divisible by 5.
If $r=4$,	$n^5 - n = 5N + 1020 = 5(N+204)$	is divisible by 5.

Thus, for all integers n, $n^5 - n$ is divisible by 5 and by 6 hence by 30;

or

$n^5 - n = n(n^4 - 1)$ is divisible by 5 since $n^{p-1} \equiv 1 \pmod{p}$ where p is a prime. $\therefore n^5 - n$ is divisible by 30.

33. $9^{x+2} - 9^x = 240$; $\therefore 9^x(81 - 1) = 240$; $\therefore 3^{2x} = 3$; $\therefore 2x = 1$; and $\therefore x = \frac{1}{2}$.

34. $x^2 + y^2 < 25$ is the set of points interior to the circle $x^2 + y^2 = 25$. Let the straight line $x + y = 1$ intersect the circle in A and B. Then all the points of the straight line segment AB, except A and B, are also interior to the circle. $\therefore$(C) is correct.

35. All the triangles so formed with A as one vertex are similar. Let h_k be the side parallel to BC at a distance $(k/8)\cdot\overline{AC}$ from A, $k = 1, 2, \ldots, 7$.

$$\therefore \frac{h_k}{(k/8)\cdot\overline{AC}} = \frac{10}{\overline{AC}};$$

$$\therefore \quad h_k = \frac{10k}{8}$$

$$S = h_1 + h_2 + \ldots + h_7 = (10/8)\,(1 + 2 + \ldots + 7) = 35.$$

Part 3

36. If $x + y = 1$, then $P = xy = x(1 - x) = -x^2 + x$. The largest value of P occurs when $x = 1/2$ (see 1950, Problem 41).

$\therefore P$ (maximum) $= 1/4$.

37. By similar triangles $\overline{MN}/x = 5/12$, or $\overline{MN} = 5x/12$. Also, $\overline{NP} = \overline{MC} = 12 - x$

$\therefore y = 12 - x + 5x/12 = (144 - 7x)/12$.

38. $10t + u - (10u + t) = 9(t - u)$. Since both t and u lie between 0 and 9, $t - u$ cannot exceed 9.

$\therefore 9(t - u)$ cannot exceed 81. The possible cubes are, therefore, 1, 8, 27 and 64, of which only 27 is divisible by 9. $\therefore 9(t - u) = 27$;

$\therefore t = u + 3$ so that the possibilities for $N = 10t + u$ are 96, 85, 74, 63, 52, 41, 30, seven in all.

39. Let t be the time each man walked. Then the distance the second man walked is given by $S = 2 + 2\frac{1}{2} + \ldots + \left(2 + \frac{t-1}{2}\right) = \frac{7t + t^2}{4}$.

$$\therefore \frac{7t + t^2}{4} + 4t = 72;$$

$\therefore t = 9$; and $\therefore$ each man walks 36 miles, so that they meet midway between M and N.

40. When the vertex is on the x-axis, $y=0$ and the roots of $-x^2+bx-8=0$ are equal since this parabola touches the x-axis at only one point.
$\therefore b^2-32=0$ and $b=\pm\sqrt{32}$.

41. There is no solution when $a(-a)-(a-1)(a+1)=0$, that is, when $a=\pm\sqrt{2}/2$ (see 1950, Problem 14);

or

solve for x, obtaining $x=(2a-1)/(2a^2-1)$, which is undefined when $2a^2-1=0$, that is, when $a=\pm\sqrt{2}/2$.

42. For integral values of n, i^n equals either i, -1, $-i$ or $+1$. When $i^n=i$, then $i^{-n}=1/i=-i/1$. $\therefore S$ can be 0, -2, or $+2$.

43. On the line $x=4$ the integral ordinates are 0, 1, 2, ..., 16; on the line $x=3$ they are 0, 1, 2, ..., 9; on $x=2$ they are 0, 1, ...,4; on $x=1$ they are 0, 1; and on $x=0$ the integral ordinate is 0.

$\therefore$ The number of lattice points is $17+10+5+2+1=35$.

44. $\angle BAD=\angle CAB-\angle CAD=\angle CAB-\angle CDA=\angle CAB-(\angle BAD+\angle B)$.

$$\therefore \quad 2\angle BAD=\angle CAB-\angle B=30°;$$
$$\therefore \quad \angle BAD=15°.$$

45. Solving for x, we have $x=y^2/(y-1)$, $y\neq 1$. Solving for y, we have $y=(x\pm\sqrt{x^2-4x})/2$.
$\therefore x(x-4)\geq 0$, so that $x\geq 4$ or $x\leq 0$ and (A) is the answer. Note that choice (B) was quickly eliminated. The values $x=4$, $y=2$ rule out choices (D) and (C). The values $x=5, y=(5\pm\sqrt{5})/2$ rule out (E).

46. The center of the circle is the point of intersection of the perpendicular bisectors of the given chords. Draw the radius to the endpoint of one of the chords and use Pythagoras' Theorem.

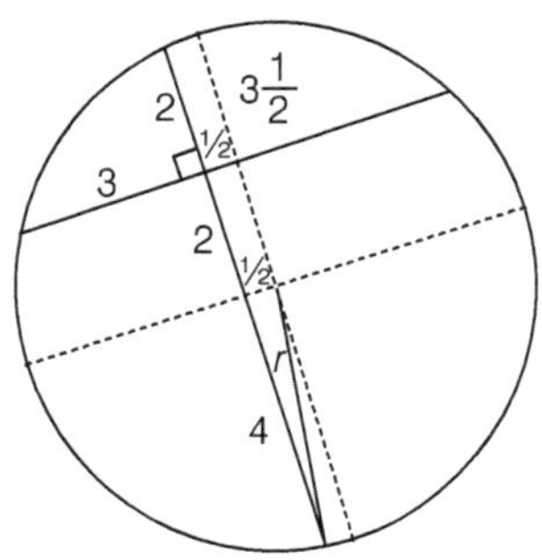

$\therefore r^2=4^2+(1/2)^2=65/4, \quad r=\sqrt{65}/2$
$d=\sqrt{65}$.

47. XY is the perpendicular bisector of AB.
$\therefore \overline{MB} = \overline{MA}$. $\measuredangle BMA$ is inscribed in a semi-circle and thus is a right angle.
$\therefore \ \measuredangle ABM = 45°$; $\therefore \widehat{AD} = 90°$; and $\therefore \overline{AD} = r\sqrt{2}$.

48. Consider first the special case when M coincides with C. Then $\overline{AM} = \overline{AC}$ and $\overline{BM} + \overline{MC} = \overline{BM} = \overline{BC} = \overline{AC}$.
$\therefore \overline{AM} = \overline{BM} + \overline{MC}$.
To prove this in general, lay off $\overline{MN}$ (on MA) equal to $\overline{MC}$. since $\measuredangle CMA = 60°$, ΔMCN is equilateral. We can show that
$\Delta ACN \cong \Delta MCB$ because $\overline{AC} = \overline{CB}$,
$\overline{CN} = \overline{CM}$, and $\measuredangle ACN = \measuredangle ACM - 60° = \measuredangle MCB$
$\therefore \overline{BM} = \overline{AN}$ and $\overline{AM} = \overline{AN} + \overline{MN} = \overline{BM} + \overline{MC}$.

49. Let x and y be the two upper segments of the non-parallel sides.
Then $3 + y + x = 9 + 6 - y + 4 - x$.
$\therefore \qquad x + y = 8$.
Since $x : y = 4 : 6 = 2 : 3$,
$2y/3 + y = 8$. $y = 24/5$ and
$y : (6 - y) = 24/5 : 6/5 = 24 : 6 = 4 : 1$

50. $A'ABB'$ is a trapezoid. Its median OO' is perpendicular to AB.

$$\overline{OO'} = \frac{1}{2}(\overline{AA'} + \overline{BB'}) = \frac{1}{2}(\overline{AG} + \overline{BG}) = \frac{1}{2}\overline{AB}.$$

$\therefore O'$ is a fixed distance from O on the perpendicular to AB, and therefore the point O' is stationary.

1958 Solutions

Part 1

1. $[2-3(-1)^{-1}]^{-1} = \left(2-\dfrac{3}{-1}\right)^{-1} = 5^{-1} = \dfrac{1}{5}.$

2. $\dfrac{y-x}{xy} = \dfrac{1}{z};\ z = \dfrac{xy}{y-x}.$

3. $\dfrac{a^{-1}b^{-1}}{a^{-3}-b^{-3}} \cdot \dfrac{a^3b^3}{a^3b^3} = \dfrac{a^2b^2}{b^3-a^3}$

4. $\dfrac{\dfrac{x+1}{x-1}+1}{\dfrac{x+1}{x-1}-1} = \dfrac{x+1+(x-1)}{x+1-(x-1)} = x.$

5. $\dfrac{(2+\sqrt{2})^2(\sqrt{2}-2)+\sqrt{2}-2+2+\sqrt{2}}{(2+\sqrt{2})(\sqrt{2}-2)} = \dfrac{-4}{-2} = 2.$

6. $\dfrac{1}{2}\left(\dfrac{x+a}{x}+\dfrac{x-a}{x}\right) = \dfrac{1}{2}\left(\dfrac{2x}{x}\right) = 1.$

7. $y = mx + b$ is satisfied by $(-1, 1)$ and $(3, 9)$, so

   $1 = -m + b$ and $9 = 3m + b$;

   $m = 2$, $b = 3$; and the equation is $y = 2x + 3$. When $y = 0$, $x = -\dfrac{3}{2}.$

8. $\sqrt{\pi^2} = \pi$, irrational $\sqrt[3]{.8} = \sqrt[3]{\dfrac{8}{10}} = \dfrac{2}{\sqrt[3]{10}}$, irrational;

   $\sqrt[4]{.00016} = \sqrt[4]{16\times 10^{-5}} = 2\cdot 10^{-1}\sqrt[4]{10^{-1}}$, irrational;

   $\sqrt[3]{-1}\cdot\sqrt{(.09)^{-1}} = -1\cdot\dfrac{1}{3}$, rational.

9. $x^2 + b^2 = a^2 - 2ax + x^2;\ x = (a^2 - b^2)/2a.$

10. $x = \dfrac{-k \pm \sqrt{k^2-4k^2}}{2}$. For x to be real, $k^2 - 4k^2 \geq 0$, an impossibility.

11. Replace $\sqrt{5-x} = x\sqrt{5-x}$ by $5 - x = x^2(5-x)$. The solution set for this equation is [5, 1, – 1] but the solution set for the original equation is {5, 1}.

or

$\sqrt{5-x} = x\sqrt{5-x}$. If $5 - x \neq 0$, divide by $\sqrt{5-x}$ and obtain $x = 1$; if $5 - x = 0$, then $x = 5$.

12. $\log P = \log s - n \log(1+k)$; $n = \dfrac{\log s - \log P}{\log(1+k)} = \dfrac{\log(s/P)}{\log(1+k)}$.

13. $\dfrac{1}{x} + \dfrac{1}{y} = \dfrac{y+x}{xy} = \dfrac{10}{20} = \dfrac{1}{2}$.

14. Set up the following table, a one-to-one correspondence between the elements of the boy set and the girl set.

Boy number 1 2 3 ... b;

Girls danced with 5 6 7 ... $b + 4$;

$\therefore g = b + 4$; $b = g - 4$;

or

by the formula for the last term of the arithmetic sequence 5, 6, 7. ..., g:

$g = 5 + (b-1)(1)$; $b = g - 4$.

15. Take a special case, the square, where all the inscribed angles are equal; and each inscribed angle $x = \dfrac{1}{2}(270°)$; $4x = 540°$.

or

see 1953, Problem 28.

16. Let r be the radius of the circle, and a, p, s the apothem, perimeter, and side of the regular hexagon.

$\pi r^2 = 100\pi$; $r = 10$; $A = \dfrac{1}{2}ap = 3as$;

$a = r = 10$;

$s = 2 \cdot 10/\sqrt{3}$;

$\therefore\ A = 3 \cdot 10 \cdot 20/\sqrt{3} = 200\sqrt{3}$.

17. $\dfrac{1}{2}\log x \geq \log 2$, $\log x \geq \log 4$, $x \geq 4$;

or

$\log x \geq \log 2\sqrt{x}$, $x \geq 2\sqrt{x}$, $\sqrt{x} \geq 2$, $x \geq 4$.

18. $A = \pi r^2$; $2A = \pi(r+n)^2$;

$2\pi r^2 = \pi(r^2 + 2rn + n^2)$;

$r = n(1+\sqrt{2})$; the negative value $1 - \sqrt{2}$ is not used.

19. $a^2 = cr$, $b^2 = cs$; $r/s = a^2/b^2 = 1/9$. (See 1956, Problem 38)

20. $4^x - 4^x$; $4^{-1} = 24$; $\frac{3}{4} \cdot 4^x = 24$; $4^x = 32$; $(2^2)^x = 2^5$; $2x = 5$; $x = \frac{5}{2}$;
$(2x)^x = 5^{5/2} = 25\sqrt{5}$.

Part 2

21. Consider the special case with A on C. The answer is obvious;

or

$A(\Delta CED) = r^2$;
$A(\Delta AOB) = \frac{1}{2} r^2$;
ratio 2 : 1.

22. When $y = 6$ $x = 4$ or -3;
when $y = -6$ $x = 0$ or 1.
Because of symmetry the particle may roll form (4, 6) to (1, − 6) or from (− 3, 6) to (0, − 6). In either case the horizontal distance is 3.

23. $y_1 = x^2 - 3$; $y_2 = (x \pm a)^2 - 3$;
$y_2 - y_1 = \pm 2ax + a^2$.

24. Let a = rate north = 30 mph, b = rate south = 120 mph. The average rate for the round trip is $\frac{2ab}{a+b}$ (see 1950, Problem 27).

$\therefore \frac{2ab}{a+b} = \frac{2 \cdot 30 \cdot 120}{30 + 120} = 48$ mph.

25. $A = \log_k x$; $\quad \therefore x = k^A$. $B = \log_5 k$; $\therefore k = 5^B$
$(\log_k x)(\log_5 k) = AB = 3$;
$x = (5^B)^A = 5^3 = 125$.

26. $s = a_1 + a_2 + \ldots\ldots + a_n$;
$s' = 5(a_1 + 20) - 20 + 5(a_2 + 20) - 20 + \ldots$
$= 5a_1 + 80 + 5a_2 + 80 + \ldots + 5a_n + 80 \quad = 5(a_1 + a_2 + \ldots + a_n) + 80n$
$= 5s + 80n$;

or

$$\sum_{i=1}^{n} a_i = s;\ \sum_{i=1}^{n} [5(a_i + 20) - 20] = 5\sum_{i=1}^{n} a_i + \sum_{i=1}^{n} 80 = 5s + 80n.$$

27. $m = \frac{\Delta y}{\Delta x} = \frac{3+3}{4-2} = \frac{(k/2) - 3}{5-4}$; $k = 12$

where $\Delta y = y_2 - y_1$, etc.

28.

Removals	Water drawn	Water left	Antifreeze drawn	Antifreeze left
First	4	12	0	4
Second	3	9	1	$3+4=7$
Third	$2\frac{1}{4}$	$6\frac{3}{4}$	$1\frac{3}{4}$	$5\frac{1}{4}+4=9\frac{1}{4}$
Fourth	$1\frac{11}{16}$	$5\frac{1}{16}$	$2\frac{5}{16}$	$6\frac{15}{16}+4=10\frac{15}{16}$

Water : $\dfrac{5\frac{1}{16}}{16}=\dfrac{81}{256}$; antifreeze $\dfrac{10\frac{15}{16}}{16}=\dfrac{175}{256}$.

29. From triangle AEC, $x+n+y+w=180°$.

From triangle BED, $m+a+w+b=180°$.

$\therefore x+y+n=a+b+m$.

30. $\dfrac{1}{x^2}+\dfrac{1}{y^2}=\dfrac{y^2+x^2}{x^2y^2}=a$; $\therefore a=\dfrac{x^2+y^2}{b^2}$ or $x^2+y^2=ab^2$;

$(x+y)^2=x^2+2xy+y^2=ab^2+2b=b(ab+2)$.

31. Denote each side by a and the base by $2b\cdot 2a+2b=32$; $a+b=16$;

$a^2-b^2=(a-b)(a+b)=(a-b)\cdot 16=64$;

$a-b=4$; $\therefore$ $b=6$.

$\therefore$ area $=8\cdot 6=48$.

32. $25s+26c=1000$; $\therefore$ $s=40-26c/25$

The only solution in positive integers is $c=25$, $s=14$.

33. Let the roots be p,q where $p=2q$; $p+q=3q=-b/a$; $9q^2=b^2/a^2$;

$pq=2q^2=c/a$. Division of $9q^2$ by $2q^2$ yields $\dfrac{9}{2}=b^2/ac$ or $2b^2=9ac$.

34. $6x+1>7-4x$; $x>\dfrac{3}{5}$; $\therefore \dfrac{3}{5}<x\le 2$.

35. There is no loss of generality by placing one vertex at the origin, O. Let the other two vertices be $A(a,c)$ and $B(b,d)$. Let D and C be the projections of A and B, respectively, on the x-axis. Therefore:

area OBA = area $OCBA$ − area OCB

= area ODA + area $DCBA$ − area OCB

$=\dfrac{1}{2}ac+\dfrac{1}{2}(b-a)(c+d)-\dfrac{1}{2}bd=\dfrac{1}{2}(bc-ad)$.

Since a,b,c,d are integers, this is rational.

36. Let h be the altitude and x, the smaller segment.
$30^2 - x^2 = h^2 = 70^2 - (80 - x)^2$; $x = 15$; $80 - x = 65$.

37. $s = \frac{n}{2}[2a + (n-1)d]$;

$a = k^2 + 1,\ n = 2k + 1,\ d = 1,$
$s = (2k + 1)(k^2 + k + 1) = 2k^3 + 3k^2 + 3k + 1$
$= k^3 + 3k^2 + 3k + 1 + k^3 = (k + 1)^3 + k^3$.

38. $s^2 - c^2 = \frac{y^2 - x^2}{r^2}$

max. val. $= \frac{y^2}{r^2}$ (when $x = 0$) $= \frac{1}{1} = 1$

min. val. $= \frac{-x^2}{r^2}$ (when $y = 0$) $= \frac{-1}{1} = -1$

or

denote by θ the angle between OP and the positive x-axis.
$\text{Sin}\,\theta = s$. $\text{Cos}\,\theta = c$. Let $s^2 - c^2 = a$.
Since $s^2 + c^2 = 1$, $2s^2 = 1 + a$; $a = 2s^2 - 1$.
Since the maximum value of s is 1, max $a = 1$, min $a = -1$.

39. If $x > 0$, $x^2 + x - 6 = 0$;

$(x - 2)(x + 3) = 0$; $x + 3 \neq 0$; $x = 2$.
If $x < 0$, $x^2 - x - 6 = 0$; $(x - 3)(x + 2) = 0$;
$x - 3 \neq 0$; $x = -2$;

or

$(|x| + 3)(|x| - 2) = 0$; $|x| + 3 \neq 0$;
$\therefore |x| = 2$, i.e., $x = 2$ or -2.

40. When $n = 1$, $a_1^2 - a_0a_2 = (-1)^1$; $a_2 = 10$.
When $n = 2$, $a_2^2 - a_1a_3 = (-1)^2$; $a_3 = 33$.

Part 3

41. $r + s = -\frac{B}{A}$; $rs = \frac{C}{A}$;

$$r^2 + 2rs + s^2 = \frac{B^2}{A^2};\ 2rs = \frac{2C}{A}$$

$$\therefore \quad r^2 + s^2 = \frac{B^2 - 2AC}{A^2}$$

$$\therefore p = -(r^2 + s^2) = \frac{2AC - B^2}{A^2}.$$

42. $\Delta AEB \sim \Delta ABD; \dfrac{AD}{12} = \dfrac{12}{8};$

$AD = 18.$

43. $16 = a^2 + \dfrac{b^2}{4}; 49 = \dfrac{a^2}{4} + b^2;$

$65 = \dfrac{5}{4}(a^2 + b^2);$

$a^2 + b^2 = c^2 = 52;$

$c = 2\sqrt{13}.$

44. Let p be the proposition "a is greater than b".

Let q be the proposition "c is greater than d".

Let r be the proposition "e is greater than f".

From the given, $p \to q$ and $\sim q \to r$.

From the Rule of the Contrapositive, $\sim q \to \sim p$ and $\sim r \to q$.

Therefore, (A), (B), (C), (D) are not valid conclusions.

45. Let $x = 10a + b$ dollars $y = 10c + d$ cents.

The face value of check is $1000a + 100b + 10c + d$ cents.

The cashed value of check is $1000c + 100d + 10a + b$ cents.

The difference is $990c + 99d - (990a + 99b) = 1786.$

$\therefore (10c + d) - (10a + b) = 18$ or $y - x = 18$. Therefore, y can equal $2x$.

46. $y = \dfrac{1}{2} \cdot \dfrac{x^2 - 2x + 2}{x - 1} = \dfrac{1}{2}\left[x - 1 + \dfrac{1}{x - 1}\right]$

The sum of a number and its reciprocal is numerically least when the number is ± 1. For $x - 1 = 1$, $x = 2$ which is excluded;

$\therefore x - 1 = -1$ and $y = -1$. All other values of x in the interval given yield values of y less than -1;

or

$$\frac{dy}{dx} = \frac{(2x-2)(2x-2) - (x^2 - 2x + 2)(2)}{(2x-2)^2} = 0;$$

$x = 0, x = 2$ (excluded).

By testing with values close to $x = 0$ to the right and to the left, we find that y is a maximum when $x = 0$.

47. $\Delta PTR \sim \Delta ATQ; \overline{PR}/\overline{AQ} = \overline{PT}/\overline{AT}$

$\overline{PT} = \overline{AT}$ ($\measuredangle PAT = \measuredangle PBS = \measuredangle APT$);

$\overline{PR} = \overline{AQ},\ \overline{PS} = \overline{QF}; \overline{PR} + \overline{PS} = \overline{AQ} + \overline{QF} = \overline{AF};$

or

$\underline{\measuredangle SBP = \measuredangle TPA = \measuredangle TAP,\ A, Q, R, P}$ are concyclic, are PR = arc AQ, $\overline{PR} = \overline{AQ}, \overline{PS} = \overline{QF}; = \overline{PR} + \overline{RS} = \overline{AF}.$

48. Let Q be the foot of the perpendicular from P to AB and let $\overline{QB} = x$. Then $(\overline{PQ})^2 = x(10-x)$.
$\overline{CP} = \sqrt{x(10-x)+(6-x)^2}$; $\overline{DP} = \sqrt{x(10-x)+(4-x)^2}$
$\overline{CP} + \overline{DP} = \sqrt{36-2x} + \sqrt{16+2x}$.
This sum is greatest when $\sqrt{36-2x} = \sqrt{16+2x}$, that is when $x = 5$. Hence (E);

or

an ellipse through A and B with C and D as foci is the locus of points P' such that $\overline{CP'} + \overline{P'D} = 10$; since this ellipse lies entirely inside the given circle (except at the points of tangency A and B), we see that

$\overline{CP} + \overline{PD} \geqq 10$ for all points P on the circle, the sign of equality holding only when P coincides with A or B.

49. $(a+b+c)^{10} = (a+b)^{10} + 10(a+b)^9c + \dfrac{10\cdot 9}{1\cdot 2}(a+b)^8c^2$
$+ \ldots + 10(a+b)c^9 + c^{10}$.

Since the expansion of $(a+b)^n$ has $n+1$ distinct terms, in extended form the first term above has 11 distinct terms, the next has 10, etc. The total number of distinct terms is

$1+2+3+\ldots+11 = \dfrac{11\cdot 12}{2} = 11\cdot 6 = 66.$

50. $\overline{DP} = x = a$; $\overline{D'P'} = y$; $\overline{PB}/\overline{P'B'} = 1/4$.

$y = \overline{D'B'} + \overline{B'P'} = 1 + \overline{B'P'}$
$= 1 + 4\overline{PB}$; $\overline{PB} = 4 - x = 4 - a$,
$y = 1 + 4(4-a) = 17 - 4a$, $x + y = 17 - 3a$,

or

$m = \dfrac{\Delta y}{\Delta x} = \dfrac{1-5}{4-3} = -4$ since for $x = 3$, $y = 5$, for $x = 4, y = 1$.

$y = -4x + b$, $y = -4x + 17$, $b = 17$.
When $x = a$, $x + y = 17 - 3a$

1959 Solutions

Part 1

1. S (old) $= 6x^2$; S (new) $= 6(1.5x)^2 = 13.50x^2$. Increase $= 7.5x^2 = \frac{5}{4}(6x^2)$.

Increase (%) $= 125$.

2. Let x be the distance from P to AB. By similar triangles

$\frac{1}{2} = \frac{(1-x)^2}{1^2}$; $\quad \therefore 1 - x = \pm\frac{1}{\sqrt{2}}$; $\quad \therefore\ x = \frac{2-\sqrt{2}}{2}$

(negative sq. root rejected).

3. There are a variety of figures satisfying the given conditions, and not falling within the classifications (A), (B), (C), (D), for example, the "kite".

4. $x + \frac{1}{3}x + \frac{1}{6}x = 78$. $9x = 468$. $\frac{1}{3}x = 17\frac{1}{3}$.

5. $(256)^{.16}(256)^{.09} = (256)^{.25} = (256)^{1/4} = 4$.

6. The converse is : If a quadrilateral is a rectangle, then it is a square.

The inverse is : If a quadrilateral is not a square, then it is not a rectangle.

Both statements are false;

or

use Venn diagrams to picture the sets mentioned.

7. $a^2 + (a+d)^2 = (a+2d)^2$; $\therefore\ a = 3d$ ($a = -d$ is excluded);

$\therefore a/d = 3/1$.

8. $x^2 - 6x + 13 = (x-3)^2 + 4$. The smallest value for this expression, 4, is obtained when $x = 3$;

or

graph $y = x^2 - 6x + 3$. The turning point, whose ordinate is 4, is a minimum point. (See 1950, Problem 41.)

9. $n = \frac{1}{2}n + \frac{1}{4}n + \frac{1}{5}n + 7$. $n = 140$.

10. Let the altitude from B to AC be h. Then since $\overline{AD} = \frac{1}{3}\overline{AB}$ the altitude from D to AE is $\frac{1}{3}$ h. Let $AE = x$. Then $\frac{1}{2}\cdot\frac{1}{3}hx = \frac{1}{2}h(3.6)$
$\therefore x = 10.8.$

11. Let $\log_2 .0625 = x$; $\therefore 2^x = .0625 = \frac{1}{2^4}$; $\therefore x = -4.$

12. Let a = the constant : $\frac{20+a}{50+a} = \frac{50+a}{100+a}$; $\therefore a = 25$; $\therefore r = \frac{5}{3}.$

13. Arithmetic mean $= \frac{\text{sum of numbers}}{\text{number of numbers}}$;
$\therefore S = 50 \times 38 = 1900$; $\therefore x = \frac{1900-45-55}{48} = 37.5.$

14. Addition, subtraction, and multiplication with even integers always yield even integers.
Note : This is a good opportunity to underscore operational restrictions imposed by the available domain.

15. $c^2 = 2ab$; $\therefore a^2 + b^2 = 2ab$; $\therefore (a-b)^2 = 0$; and $\therefore a = b.$
Since the right triangle is isosceles, one of the acute angles is 45°.

16. $\frac{(x-2)(x-1)}{(x-3)(x-2)}\cdot\frac{(x-3)(x-4)}{(x-4)(x-1)} = 1.$

17. $y = a + \frac{b}{x}$, $1 = a - b$, $5 = a - \frac{1}{5}b.$
$\therefore \quad b = 5$ and $a = 6.$

18. The sum of the first n positive integers is $s = \frac{n(n+1)}{2}$. Since A.M. $= \frac{s}{n}$, the correct choice is (E).

19.

	Weights used	No. of weighings possible
1.	singly	3
2.	two at a time (same pan)	3
3.	three at a time (same pan)	1
4.	two at a time (diff. pans)	3
5.	three at a time (diff. pans)	3
	Total	13

20. $x = \frac{Ky}{Z^2}$, $10 = \frac{K(4)}{14^2}$; $\therefore K = \frac{10\cdot 14^2}{4}$; and $\therefore x = \frac{\frac{10\cdot 14^2}{4}\cdot 16}{7^2} = 160.$

Part 2

21. $R = \frac{2}{3}$ altitude $= \frac{2}{3} \cdot \frac{1}{2} \cdot \frac{p}{3}\sqrt{3} = \frac{p}{3\sqrt{3}}$; $\therefore\ A = \frac{\pi p^2}{27}$.

22. The median of a trapezoid goes through the midpoints of the diagonals. Let x be the length of the shorter base.

$\therefore$ length of median $= \frac{x}{2} + 3 + \frac{x}{2}$; $\therefore \frac{1}{2}(x + 97) = \frac{x}{2} + 3 + \frac{x}{2}$; $\therefore\ x = 91$.

23. $\log_{10}(a^2 - 15a) = 2$; $\therefore\ 10^2 = a^2 - 15a$; and $\therefore\ a = 20$ or $a = -5$.

24. $\dfrac{\frac{m^2}{100} + x}{m + x} = \dfrac{2m}{100}$; $\therefore\ x = \dfrac{m^2}{100 - 2m}$.

25. Let AB be a straight line segment with its midpoint at 3, and its right end at B. Each of the two intervals from A to 3 and from 3 to B is less than 4. Hence B is to the left of 7 and A is to the right of -1;

or

if $x \leq 3$, $|3 - x| < 4$ means $3 - x < 4$, $\therefore -1 < x$, if $x > 3$, $|3 - x| < 4$ means $x - 3 < 4$, $\therefore\ x < 7$.

26. $x^2 + x^2 = 2$, $x = 1$; $\therefore\ \overline{DF} = \frac{\sqrt{2}}{2}$.

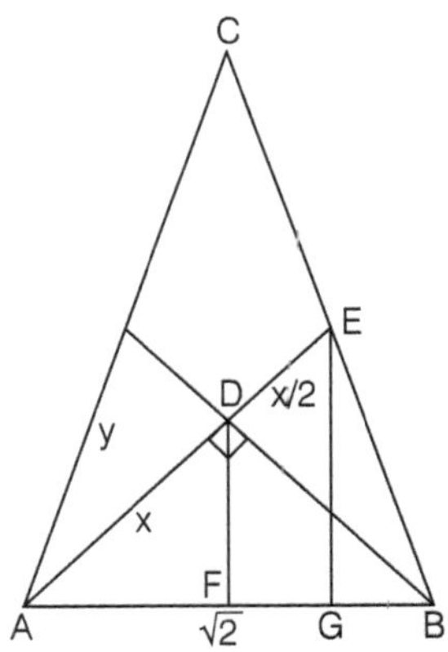

Draw $EG \perp AB$. $\dfrac{\overline{AD}}{\overline{AE}} = \dfrac{2}{3} = \dfrac{\overline{DF}}{\overline{EG}}$;

$\therefore \overline{EG} = \frac{3}{2}\overline{DF} = \frac{3\sqrt{2}}{4}$.

Altitude $CDF = 2\overline{EG} = \frac{3\sqrt{2}}{2}$,

$\therefore$ Area $= \frac{1}{2} \cdot \sqrt{2} \cdot \frac{3\sqrt{2}}{2} = \frac{3}{2}$

or

$x^2 + x^2 = 2$, $x = 1$; $y^2 = x^2 + \left(\frac{x}{2}\right)^2 = 1 + \frac{1}{4} = \frac{5}{4}$.

Altitude $CDF = \sqrt{(2y)^2 - \left(\frac{\sqrt{2}}{2}\right)^2} = \sqrt{4 \cdot \frac{5}{4} - \frac{1}{2}}$

$$= \sqrt{\frac{9}{2}} = \frac{3}{\sqrt{2}} = \frac{3\sqrt{2}}{2}, \text{ etc.}$$

27. The sum of the roots is $1/i = -i$.

28. The bisector of an angle of a triangle divides the opposite sides into segments proportional to the other two sides (see 1953, Problem 28).
$\therefore \frac{\overline{AM}}{\overline{MB}} = \frac{b}{a}$ and $\frac{\overline{CL}}{\overline{LB}} = \frac{b}{c}$. Since $\frac{c}{a} \cdot \frac{b}{c} = \frac{b}{a}$, $k = \frac{c}{a}$.

29. $\dfrac{15 + \frac{1}{3}(x-20)}{x} = \frac{1}{2}$; $\therefore$ $x = 50$.

30. Let $x = B$'s time in seconds. Letting l represent the length of the track, we have $\frac{l}{40}(15) + \frac{l}{x}(15) = 1$. $\therefore$ $x = 24$.

31. Let $2s$ be the side of the smaller square. Then $\frac{r-s}{2s} = \frac{2s}{r+s}$; $\therefore r = \sqrt{50}$.

Let S be the side of the larger square. Then
$S = r\sqrt{2} = 10$; $\therefore A = 100$.

32. The point A, the point of tengency, and the center of the circle, determine a right triangle with one side l, another side r, and the hypotenuse $x + r$, where x is the shortest distance from A to the circle.

$$\left(\frac{4r}{3}\right)^2 + r^2 = (x + r)^2; \;\therefore\; x = \frac{2r}{3} = \frac{l}{2}.$$

33. $\frac{1}{3}, \frac{1}{4}, \frac{1}{6}, \frac{1}{12}, 0$ with $d = -\frac{1}{12}$.

$\therefore$ the terms of the H.P. are 3, 4, 6, 12 only (since 0 has no reciprocal).
$\therefore S_4 = 25$.

34. $x^2 - 3x + 1 = 0$; $\therefore rs = 1$ and $r + s = 3$; $\therefore (r+s)^2 = r^2 + 2rs + s^2 = 9$; and $\therefore$ $r^2 + s^2 = 7$.

35. Solving for x we have $x(-m+n) = -mn + n^2$; $\therefore x = n$.

36. Let the triangle be ABC, with $\overline{AB} = 80$, $\overline{BC} = a$, $\overline{CA} = b = 90 - a$, $\measuredangle B = 60°$. Let CD be the altitude to AB. Let $x = \overline{BD}$.
$\therefore \overline{CD} = \sqrt{3}x$, $a = 2x$, $b = 90 - 2x$;
$\therefore 3x^2 + (80 - x)^2 = (90 - 2x)^2$; $\therefore$ $x = \frac{17}{2}$; and $\therefore$ $a = 17$, $b = 73$.

37. $$\left(1-\frac{1}{3}\right)\left(1-\frac{1}{4}\right)\left(1-\frac{1}{5}\right)\cdots\left(1-\frac{1}{n}\right)$$
$$= \frac{2}{3}\cdot\frac{3}{4}\cdot\frac{4}{5}\cdots\frac{n-2}{n-1}\cdot\frac{n-1}{n} = \frac{2}{n}.$$

38. $4x + \sqrt{2x} = 1$, or $4x - 1 = -\sqrt{2x}$. $\therefore 16x^2 - 10x + 1 = 0$;

$\therefore x = \frac{1}{8}$. ($x = \frac{1}{2}$ does not satisfy the original equation.);

or

let $y = \sqrt{x}$ $\therefore 4y^2 + \sqrt{2}y - 1 = 0$ and $y = \frac{2\sqrt{2}}{8}$, or

$y = \frac{-4\sqrt{2}}{8}$. Thus $x = y^2 = \frac{1}{8}, \frac{1}{2}$ (Reject).

39. $x + x^2 + x^3 + \ldots + x^9$; $S_1 = \frac{x(1-x^9)}{1-x} = \frac{x^{10}-x}{x-1}$.

$a + 2a + 3a + \ldots + 9a$; $S_2 = \frac{9}{2}(a + 9a) = 45a$; $S = S_1 + S_2$.

40. Let G be a point on EC so that $\overline{FE} = \overline{EG}$. Connect D with G. Then $FDGB$ is a parallelogram. $\therefore \overline{DG} = 5$, $\overline{AF} = 10$, $\overline{AB} = 15$.

Part 3

41. $x^2 + (x-4)^2 = (x+4)^2$; $\therefore x = 16, 0$.

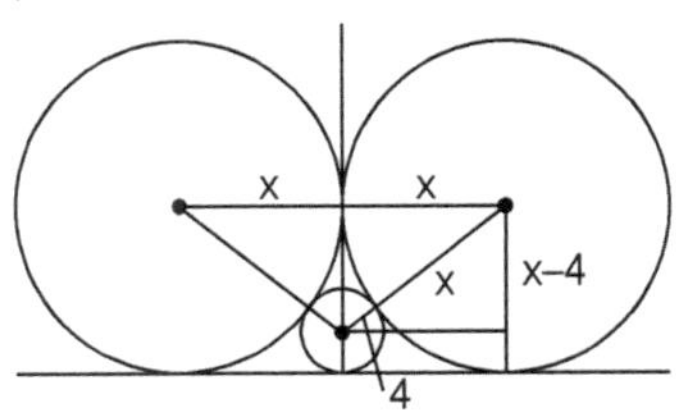

42. Represent a, b, c in terms of their prime factors. Then D is the product of all the common prime factors, each factor taken as often as it appears the least number of times in a or b or c. M is the product of all the non-common prime factors, each factor taken as often as it appears the greatest number of times in a or b or c.

Therefore, MD may be less than abc, but it cannot exceed abc. Obviourly, MD equals abc when there are no common factors.

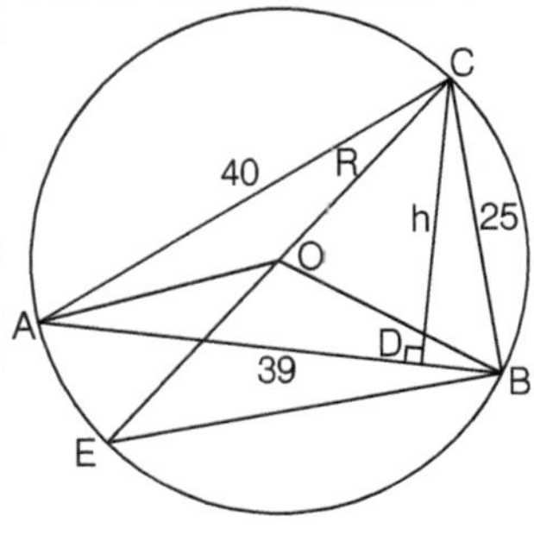

43. Let h be the altitude to side AB. Any other altitude can be used equally well. Then right triangle ACD is similar to right triangle ECB because angle A and angle E are equal. Hence $2R/25 = 40/h$. To find h, we use the area:

Area of $ABC = \frac{1}{2} h \cdot 39$

$$= \sqrt{s(s-a)(s-b)(s-c)}$$
$$= \sqrt{(52)(12)(13)(27)}$$
$$= 4\cdot 9\cdot 13$$
$$h = \frac{2\cdot 4\cdot 9\cdot 13}{39} = 24.$$
$$2R = \frac{25\cdot 40}{24} = \frac{125}{3}.$$

44. Let the roots be $1+m$ and $1+n$ with m and n both positive.

$\therefore 1+m+1+n=-b$ and $(1+m)(1+n)=c.$

$\therefore s=b+c+1=mn>0.$

45. Since $\log_a b = \dfrac{\log_c b}{\log_c a}$, we have, with base x,

$$\frac{\log x}{\log 3}\cdot\frac{\log 2x}{\log x}\cdot\frac{\log y}{\log 2x}=2$$

$\log y = 2\log 3 = \log 9;\ \therefore\ \ y=9.$

46.

	rainy mornings	non-rainy mornings
rainy afternoons	a	b
non-rainy afternoons	c	e

$d=a+b+c+e,\ a+b+c=7,\ a=0,\ c+e=5,\ b+e=6;$

$\therefore e=2$, and $\therefore d=9.$

47.

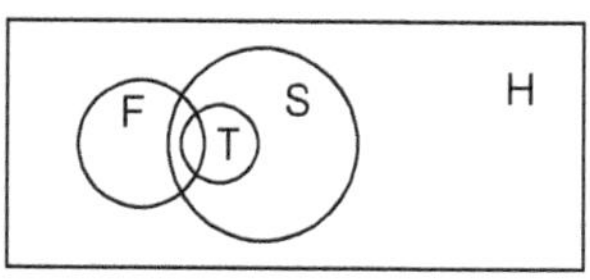

48. For $h=3$ we can have $1x^2, 1x^1+1, 1x^1-1, 2x^1, 3x^0$ †.

† The number h is known as the height of the polynomial. It is utilized in proving that the number of polynomials with integer coefficients is a countable infinity, the lowest one in the scale of infinities.

49. Combine the terms in threes, to get the geometric series.

$$\frac{1}{4}+\frac{1}{32}+\frac{1}{256}+\dots;\quad \therefore S = \frac{\frac{1}{4}}{1-\frac{1}{8}} = \frac{2}{7};$$

or

rearrange the terms into three series:

$$1+\frac{1}{8}+\frac{1}{64}+\ldots,\ -\frac{1}{2}-\frac{1}{16}-\frac{1}{128}-\ldots,\ -\frac{1}{4}-\frac{1}{32}-\frac{1}{256}-\ldots.$$

$$S_1=\frac{1}{1-\frac{1}{8}}=\frac{8}{7};\quad S_2=\frac{-\frac{1}{2}}{1-\frac{1}{8}}=-\frac{4}{7};\ S_3=\frac{-\frac{1}{4}}{1-\frac{1}{8}}=-\frac{2}{7}.$$

$$\therefore\qquad S=\frac{2}{7}.$$

50. This problem may be interpreted as a miniature finite geometry of 4 lines and 6 points, so that each pair of lines has only 1 point in common, and each pair of points has only 1 line in common;

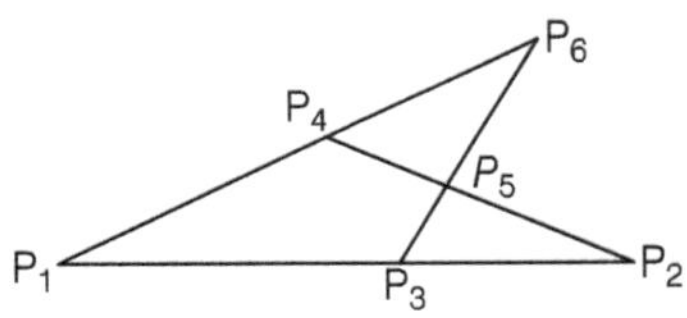

or

$C(4,2)=\frac{4!}{2!2!}=6$, displayed as follows:

Committees	*A*	*B*	*C*	*D*
Members	*a*	*b*	*d*	*f*
	c	*e*	*a*	*e*
	b	*d*	*f*	*c*

1960 Solutions

Part 1

1. Substituting 2 for x, we have $2^3 + h\cdot 2 + 10 = 0$. $\therefore h = -9$.

2. For the six strokes there are five equal intervals so that the time interval between successive strokes is one second. For twelve strokes, then, eleven seconds are required.

3. 40% of \$10,000 is \$4,000; 36% of \$10,000 is \$3,600;
   4% of (\$10,000 – \$3,600) is \$256; \$3,600 + \$256 = \$3,856;
   $\therefore$ the difference is \$4,000 – \$3,856 = \$144,

   or

   two successive discounts of 36% and 4% are equivalent to one discount of 38.56% (see 1950, Problem 22).

4. The triangle is equiangular with side 4.
   $A = s^2\sqrt{3}/4 = 4^2\sqrt{3}/4 = 4\sqrt{3}$.

5. Substitute $y^2 = 9$ into $x^2 + y^2 = 9$ to obtain $x^2 + 9 = 9$, so that $x = 0$. Therefore, the common solutions, and, hence, the intersection points are $(0, 3)$ and $(0, -3)$;

   or

   $x^2 + y^2 = 9$ represents a circle with radius 3 and centered at the origin. $y^2 = 9$ represents the pair of lines $y = +3$ and $y = -3$, tangent to the circle at $(0, 3)$ and $(0, -3)$.

6. Since $100 = \pi d$, $d = 100/\pi$. Let s be the side of the inscribed square.
   $s = d/\sqrt{2} = 100/\pi\sqrt{2} = 50\sqrt{2}/\pi$.

7. Area II/Area I $= \pi R^2/\pi r^2 = (2r)^2/r^2 = 4$;
   $\therefore$ Area II = 4 Area I $= 4\cdot 4 = 16$.

8. Let $N = 2.52525\ldots$, $100N = 252.52525\ldots$ Subtracting the first equality from the second, we obtain $99N = 250$, $\therefore N = 250/99$.

   Since this fraction is already in lowest terms, we have $250 + 99 = 349$.

9. $\dfrac{a^2+b^2-c^2+2ab}{a^2+c^2-b^2+2ac}=\dfrac{(a+b)^2-c^2}{(a+c)^2-b^2}$

$=\dfrac{(a+b+c)(a+b-c)}{(a+c+b)(a+c-b)}=\dfrac{a+b-c}{a+c-b}$ with $(a+c)^2\neq b^2$.

10. To negate the given statement we say: It is false that all men are good drivers, that is, at least one man is a bad driver.

11. The product of the roots, here, is equal to $2k^2-1$.

$\therefore 2k^2-1=7,\ k^2=4.$

The discriminant $=(-3k)^2-4\cdot1\cdot(2k^2-1)=k^2+4=8$; therefore the roots are irrational.

12. The locus is a circle, centered at the given point, having a radius equal to that of the given circles.

13. The given lines intersect pairwise in the three points $A(0,2)$, $B(4/3,-2)$ and $C(-4/3,-2)$. These three points are the vertices of a triangle with $\overline{AC}=\overline{AB}$, each of length $4\sqrt{10}/3$, and $\overline{BC}=8/3$, so that (B) is the correct choice.

14. $3x-5+a=bx+1,\ 3x-bx=6-a,\ x=(6-a)/(3-b)$ if $b\neq3$.

15. Since the triangles are similar, we have $A/a=P/p=R/r=\sqrt{K}/\sqrt{k}$ always, so that (B) is the correct choice.

16. $69=2\cdot5^2+3\cdot5+4\cdot1=234_5$ (that is, 234 in the base 5 system). The correct choice is, therefore, (C). (See also 1956, Problem 31.)

17. We have $N=8\cdot10^8\cdot x^{-3/2}$ with $N=800$; solve for $x: 800=8\cdot10^8\cdot x^{-3/2}$, $1=10^6\cdot x^{-3/2},\ x^{3/2}=10^6,\ x^{1/2}=10^2,\ x=10^4$.

18. $3^{x+y}=81=3^4,\ \therefore x+y=4,\ 81^{x-y}=3=81^{1/4}$

$\therefore x-y=\dfrac{1}{4},$

$\therefore\ x=2+\dfrac{1}{8},\ y=2-\dfrac{1}{8}$ and the correct choice is (E).

19. For (A), we have $n+n+1+n+2=3n+3=46$, an impossibility for integral n.

For (B), we have $2n+2n+2+2n+4=6n+6=46$, an impossible for integral n.

For (C), we have $n+n+1+n+2+n+3=4n+6=46$, solvable for integral n.

For (D), we have $2n+2n+2+2n+4+2n+6=8n+12=46$, an impossibility for integral n.

For (E), we have $2n+1+2n+3+2n+5+2n+7,=8n+16=46$, an impossibility for integral n.

20. $\left(\dfrac{x^2}{2}-\dfrac{2}{x}\right)^8 = \dfrac{1}{(2x)^8}(x^3-4)^8$

$= \dfrac{1}{(2x)^8}\left[(x^3)^8 + 8(x^3)^7(-4) + \dfrac{8\cdot 7}{1\cdot 2}(x^3)^6(-4)^2 + \dfrac{8\cdot 7\cdot 6}{1\cdot 2\cdot 3}(x^3)^5(-4)^3 + \ldots\right]$

the required term is $\dfrac{1}{2^8\cdot x^8}\cdot\dfrac{8\cdot 7\cdot 6}{1\cdot 2\cdot 3}(x^3)^5(-4)^3$; the coefficient is -14;

or

$\dfrac{1}{2^8x^8}C(8,5)(x^3)^5\ (-4)^3$

$= \dfrac{1}{2^8x^8}\cdot\dfrac{8\cdot 7\cdot 6}{1\cdot 2\cdot 3}(-4)^3x^7 = -14x^7$;

$\therefore$ The required coefficient is -14.

Part 2

21. Let s be a side of square I; S, a side of square II. Area I $= s^2$, Area II $= S^2$. Since $S^2 = 2s^2$; $S = s\sqrt{2}$. But $s = d/\sqrt{2} = (a+b)/\sqrt{2}$,

$\therefore S = [(a+b)/\sqrt{2}]\sqrt{2} = a+b$;

$\therefore$ the perimeter of II $= 4S = 4(a+b)$.

22. $(x+m)^2-(x+n)^2=(m-n)^2$. Factor the left side of the equation as the difference of two squares.

$(x+m-x-n)(x+m+x+n)=(m-n)^2$

$(m-n)(2x+m+n)=(m-n)^2$

Since $m\neq n$, $2x+m+n=m-n$, $x=-n$, so that (A) is the correct choice.

23. $V=\pi R^2H$, $\pi(R+x)^2H=\pi R^2(H+x)$;

$R^2H+2RxH+x^2H=R^2H+R^2x$;

$2RxH+x^2H=R^2x$; since $x\neq 0$, $2RH+xH=R^2$;

$x=\dfrac{R^2-2RH}{H}$

For $R=8$ and $H=3$, $x=\dfrac{16}{3}$, so that (C) is the correct choice.

24. $\log_{2x}216=x$, $(2x)^x=216$, $2^x\cdot x^x=2^3\cdot 3^3$.

an obvious solution to this equation is $x=3$, so that (A) is the correct choice.

25. $m=2r+1$ where $r=0,\pm 1,\pm 2\ldots$;

$n=2s+1$ where $s=0,\pm 1,\pm 2,\ldots$;

$m^2-n^2=4r^2+4r+1-4s^2-4s-1$

$= 4(r-s)(r+s+1)$, a number certainly divisible by 4.

If r and s are both even or both odd, $r-s$ is divisible by 2, and $r+s+1$ is not.

If r and s are one even and one odd, then $r+s+1$ is divisible by 2, and $r-s$ is not.

Thus $m^2 - n^2$ is divisible by $4 \cdot 2 = 8$.

26. The numerical value of $|5-x|$ must be less than 6 for the inequality to hold. Therefore, x must be greater than -1, and less than 11;

or

$$\left|\frac{5-x}{3}\right| < 2, \; -2 < \frac{5-x}{3} < 2, \; -6 < 5-x < 6;$$

$\therefore -6 < 5-x, \; x < 11$ and $5-x < 6, x > -1$; $\therefore -1 < x < 11$.

27. Since each interior angle is $7\frac{1}{2}$ times its associated exterior angle,

$$S = \left(7\frac{1}{2}\right)(\text{ sum of the exterior angles}) = \frac{15}{2} \cdot 360 = 2700.$$

Since $S = (n-2)\,180$, $(n-2)\,180 = 2700$, $n = 17$.A 17-sided polygon that is equiangular may or may not be equilateral, and, hence, regular.

28. A formal solution of $x - \dfrac{7}{x-3} = 3 - \dfrac{7}{x-3}$ yields a repeated value of 3 for x. But $x = 3$ makes the expression $\dfrac{7}{x-3}$ meaningless. The correct choice is, therefore, (B).

29. $5a + b > 51$.

$3a - b = 21$,

$\therefore 8a > 72$, $a > 9$, $3a > 27$.

Since $b = 3a - 21$, $b > 6$.

30. The locus of points equidistant from the coordinate axes is the pair of lines $y = x$ and $y = -x$.

Solve simultaneously $3x + 5y = 15$ and $y = x$ to obtain $x = y = \dfrac{15}{8}$. The point $\left(\dfrac{15}{8}, \dfrac{15}{8}\right)$ in quadrant I, therefore, satisfies the required condition.

Solve simultaneously $3x + 5y = 15$ and $y = -x$ to obtain $x = -15/2$, $y = 15/2$. The point $(-15/2, 15/2)$ in quadrant II, therefore, satisfies the required condition.

There are no other solutions to these pairs of equations, and, therefore, no other points satisfying the required condition.

Part 3

31. Let the other factor be $x^2 + ax + b$. Then

$(x^2 + 2x + 5)\,(x^2 + ax + b)$

$\equiv x^4 + x^3\,(2 + a) + x^2(5 + b + 2a) + x(5a + 2b) + 5b$

$\equiv x^4 + px^2 + q.$

Macth the coefficients of like powers of x:

For $x^3 : 2 + a = 0; \therefore a = -2$, For $x : 5a + 2b = 0; \therefore b = 5$,

For $x^2 : 5 + b + 2a = p\,; \ \therefore p = 6$, For $x^0 : 5b = q; \therefore q = 25;$

or

let $y = x^2$ so that $x^4 + px^2 + q = y^2 + py + q$. Let the roots of $y^2 + py + q = 0$ be r^2 and s^2. Since $y = x^2$ the roots of $x^4 + px^2 + q = 0$ must be $\pm r, \pm s$.

Now $x^2 + 2x + 5$ is a factor of $x^4 + px^2 + q$; consequently, one pair of roots, say r and s, must satisfy the equation $x^2 + 2x + 5 = 0$. It follows that $-r$ and $-s$ must satisfy the equation $x^2 - 2x + 5 = 0$. Therefore, the other factor must be $x^2 - 2x + 5$.

$\therefore (x^2 + 2x + 5)\,(x^2 - 2x + 5) \equiv x^4 + 6x^2 + 25 \equiv x^4 + px^2 + q.$

$\therefore p = 6,\ q = 25;$

or

$$
\begin{array}{r|l}
 & x^2 \quad - \quad 2x \quad + (p-1) \\
\hline
x^2 + 2x + 5 & x^4 \quad + \quad px^2 \quad + q \\
 & \underline{x^4 + 2x^3 + 5x^2} \\
 & -2x^3 + (p-5)x^2 \\
 & \underline{-2x^3 - 4x^2 - 10x} \\
 & (p-1)x^2 + 10x + q \\
 & \underline{(p-1)x^2 + 2(p-1)x + 5(p-1)} \\
 & (12 - 2p)x + (q - 5p + 5).
\end{array}
$$

Since the remainder must be zero (why?), $12 - 2p = 0$, $p = 6$ and $q - 5p + 5 = 0$, $q = 25$.

32. Since AB is tangent to the circle, we have $\overline{AD}/\overline{AB} = \overline{AB}/\overline{AE}$, $\overline{AB}^2 = \overline{AD}\cdot\overline{AE}$. But $\overline{AE} = \overline{AD} + 2r = \overline{AD} + \overline{AB}$;

$\therefore \overline{AB}^2 = \overline{AD}(\overline{AD} + \overline{AB}) = \overline{AD}^2 + \overline{AD}\cdot\overline{AB}.$

$\therefore \overline{AD}^2 = \overline{AB}^2 - \overline{AD}\cdot\overline{AB} = \overline{AB}(\overline{AB} - \overline{AD}).$

Since $\overline{AP} = \overline{AD}$, $\overline{AP}^2 = \overline{AB}(\overline{AB} - \overline{AP}) = \overline{AB}\cdot\overline{PB}.$

33. All those numbers $P+n$ of the sequence $P+2$, $P+3$, ..., $P+59$ for which n is prime are divisible by n because n already occurs as a factor in P. In the remaining members, n is composite and hence can be factored into primes that are smaller than 59. Hence all terms are divisible by n.

$\therefore$ all members of the sequence are composite.

34. We graph the position of each swimmer with respect to time [this graph ends with $t=3$ (minutes)].

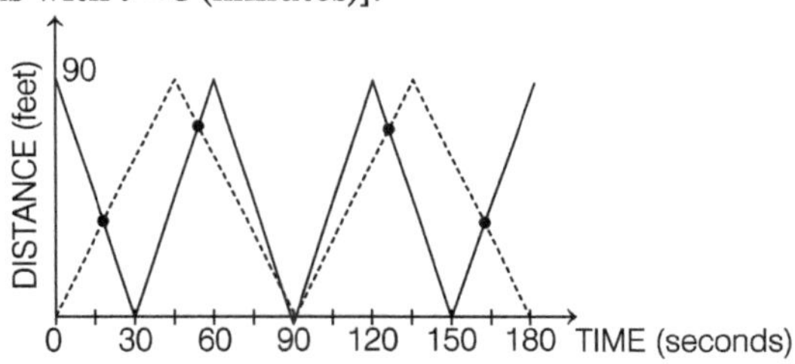

At the end of 3 minutes they are back to their original positions, so that in 12 minutes this cycle is repeated four times. Since there are five meetings in this cycle, the total number of meetings is $4 \cdot 5 = 20$.

The problem may be viewed as one in periodic phenomena; the period for the faster swimmer is 60 seconds, while the period for the slower swimmer is 90 seconds (the period is the time-interval necessary for the swimmer to return to his starting point). The common period is 180 seconds or 3 minutes.

35. $m/t = t/n$, $mn = t^2$. Since $m+n=10, t^2 = m(10-m)$.

$\therefore t = \sqrt{m(10-m)}$.

Since m, $10-m$, and t are integers, we have $t=3$ when $m=1$ and $t=4$ when $m=2$. The remaining integral values of m, namely $m = 3, 4, 6, 7$, yield non-integral values of t. The value $m=5$ cannot be used, since m and n are unequal.

36. $s_1 = \dfrac{n}{2}[2a + (n-1)d]$,

$s_2 = \dfrac{2n}{2}[2a + (2n-1)d]$, $s_3 = \dfrac{3n}{2}[2a+(3n-1)d]$.

$R = s_3 - s_2 - s_1 = \dfrac{n}{2}[6a + 9nd - 3d - 4a - 4nd + 2d - 2a - nd + d]$

$= 2n^2 d$,

so that (B) is the correct choice.

37. Designate the base of the rectangle by y. Then, because of similar triangles, $\dfrac{h-x}{y} = \dfrac{h}{b}$, $y = \dfrac{b}{h}(h-x)$. $\therefore$ area $= xy = \dfrac{bx}{h}(h-x)$.

38. $b + 60 = a + B$

$a + 60 = c + C$

$\therefore b - a = a - c + B - C.$

Since $B = C$, we have $b + c = 2a$, $a = \dfrac{b+c}{2}$.

39. $\dfrac{a+b}{a} = \dfrac{b}{a+b}$, $a^2 + 2ab + b^2 = ab$, $a^2 + ab + b^2 = 0$.

$\therefore a = \dfrac{-b \pm \sqrt{b^2 - 4b^2}}{2} = \dfrac{-b \pm ib\sqrt{3}}{2}$.

If b is real, a is not real. If b is not real, a may be real or not real.

40. Since CD trisects right angle $C, \measuredangle BCD = 30°$ and $\measuredangle DCA = 60°$ and $\measuredangle CDE = 30°$,

so that ΔDEC is a $30° - 60° - 90°$ triangle.

Let $\overline{EC} = x$.

$\therefore \overline{DE} = x\sqrt{3}$ and $\overline{DC} = 2x$.

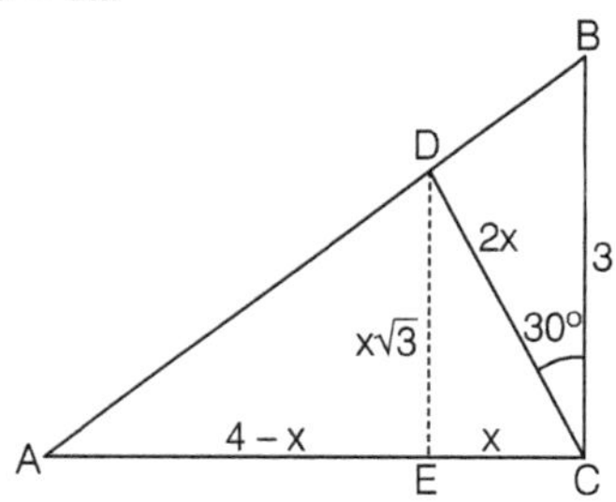

$\dfrac{4-x}{x\sqrt{3}} = \dfrac{4}{3}$, $x = \dfrac{12}{3+4\sqrt{3}}$,

$2x = \dfrac{24}{3+4\sqrt{3}} = \dfrac{32\sqrt{3}-24}{13}$;

or

Area of $\Delta BCD = \dfrac{1}{2} \cdot 3 \cdot 2x \cdot \sin 30° = \dfrac{3x}{2}$.

Area of $\Delta ACD = \dfrac{1}{2} \cdot 4 \cdot 2x \cdot \sin 60° = 2\sqrt{3}x$.

Area of ΔBCD + Area of ΔACD = Area of ΔACB.

$\therefore \dfrac{3}{2}x + 2\sqrt{3}x = 6$, $x = \dfrac{12}{3+4\sqrt{3}}$

$= \dfrac{16\sqrt{3}-12}{13}$, $2x = \dfrac{32\sqrt{3}-24}{13}$.

Classification of Problems

The classification of these contest problems is not a simple task; their content is so varied and their solution-possibilities so diverse that it is difficult to pigeonhole them into a few categories. Moreover, no matter which main topics one selects, there are always borderline cases that must be cross-indexed.

Arithmetic, *Algebra*, *Geometry*, and *Miscellaneous Topics* constitute the primary groupings. Subheadings provide a somewhat more refined classification of the problems in the four main categories, the last of which has but two subheadings, Logic and Sets. Some of the problems in these two divisions are also listed elsewhere, but others could not satisfactorily be classified in any other way.

Under *Arithmetic* are subsumed approximate-number and number-theoretic problems, and problems dealing with the arithmetic mean, with percentage, and with profit-and-loss; these are also cross-indexed with appropriate subheadings under *Algebra*. The majority of the problems listed under *Geometry* concern Euclidean geometry of the plane; the relatively few problems in coordinate geometry and solid geometry are also included here.

The number preceding the semicolon refers to the last two digits of the examination year and the number following the semicolon refers to the problem in that examination. For example, 59; 13 means Problem 13 in the 1959 examination.

	Arithmetic
Approximate numbers	**51**; 7 **53**; 29
Mean	**50**; 5 **51**; 43 **52**; 2, 45 **53**; 45 **55**; 3, 38 **57**; 20 **58**; 6, 24, 26 **59**; 13, 18;
Number theory	**50**; 7 **51**; 15, 19, 37, 45 **54**; 4 **56**; 31, 34 **57**; 19, 24, 32, 38 **59**; 19, 42 **60**;8, 16, 19, 25, 33
Percent	**50**; 8, 22, 36 **51**; 1, 8, 13 **52**; 16, 34 **53**; 2, 19, 43 **54**; 7, 33 **55**;7, 22 **56**; 13, 45 **59**; 1, 24 **60**;3
Profit and loss	**50**; 36 **51**; 5 **52**; 14, 17 **53**;1, 16, 30 **54**; 23 **55**; 26 **56**;2

Algebra

Binomial expansion see Polynomial expansion
Discriminants see Roots
Equations
Cubic (and higher) **50**; 13 **53**; 4 **54**; 20, 41 **60**; 1
Indeterminate **56**; 6 **57**; 45 **58**; 32 **60**; 19
Linear **50**; 1, 2, 19, 23, 27, 28, 29, 31, 50
51; 30, 38, 39, 50 **52**; 4, 9, 10, 32
53; 6, 9, 10, 17, 31 **54**; 7, 30, 36, 48, 50
55; 2, 4, 6, 7, 26, 33, 35, 40, 41, 47, 50
56; 2, 11, 16, 19, 22, 28, 40, 41, 46, 47
57; 11, 31, 45, **58**; 2, 34, 40, 45, 50
59; 9, 17, 29, 30, 46 **60**; 2, 14, 22, 29, 34
Quadratic **50**; 38, 42 **51**; 31 **52**; 6, 23 **53**; 44
54; 2, 24, 25, 44, 47 **55**; 18, 19, 27, 30, 32
56; 7, 14, 15, 23 **57**; 2, 27 **58**; 9, 10, 33, 39, 41
59; 23, 27, 32, 34, 35, 44 **60**; 23, 28, 39
Radical **50**; 24 **51**; 42 **53**; 23 **54**; 21
55; 12, 30, 42 **56**; 42 **57**; 22 **58**; 11
59; 38 **60**; 17
Systems **50**; 6, 14, 30 **51**; 35, 44 **52**; 44, 47, 48
53; 8 **54**; 12 **55**; 19, 38, 43, 46 **56**; 48
57; 41 **60**; 5, 18
Exponents **51**; 20, 24, 35 **52**; 7, 47 **54**; 6, 34
55; 1, 13 **56**; 1, 3, 8, 9, 18 **57**; 9, 12, 33
58; 1, 3, 20 **59**; 5 **60**; 17, 18
Extreme value **50**; 9, 41 **51**; 32, 43
52; 13 **55**; 39 **57**; 10, 36
58; 17, 38, 46, 48 **59**; 8, 32
Factors **50**; 15 **51**; 8 **52**; 3, 26 **53**; 3, 18 **54**; 32, 49 **55**; 25
Fractions **50**; 4, 40 **51**; 40 **52**; 7, **53**; 7 **54**; 22
55; 13 **56**; 11, 12, 17 **57**; 3 **59**; 16 **60**; 9, 31
Functions
Linear **55**; 10, 23, 37 **58**; 4
Notation **52**; 11 **53**; 35, 38 **54**;16
Polynomial **54**; 17 **59**; 48
Quadratic **52**; 13, 28, 40 **55**; 24 **56**; 41 **57**; 15, 40 **58**; 23, 30, 38 **59**; 8 **60**; 25, 31

Graphs	**50**; 44 **51**; 16, 33 **54**; 42, 45 **55**; 8, 28, 46, 49, **56**; 21, 29 **57**; 10, 16, 23, 34, 36, 40, 41, 43 **58**; 7, 22, 27, 46, 50 **59**; 50 **60**; 5
Inequalities	**51**; 21, 43 **53**; 47 **54**; 18 **55**; 40 **56**; 44 **57**; 29, 34 **58**; 10, 34 **59**; 25, 35, 44 **60**; 26, 29
Logarithms	**50**; 25, 26, 37 **51**; 22, 34, 45 **52**; 18 **53**; 5, 21, 22, 39, 47 **54**; 15, 38 **55**; 17 **56**; 20 **57**; 5, 28 **58**; 12, 17, 25 **59**; 11, 23, 45 **60**; 24
Numbers	
Complex	**57**; 42 **59**; 27 **60**; 39
Real	**52**; 1 **54**; 34 **56**; 33 **57**; 13 **58**; 5, 8
Reciprocal	**58**; 13, 46
Operations	**50**; 10, 18, 20 **52**; 11, 28, 35, 36, 42 **53**; 20, 24, 36 **54**; 1, 14 **55**; 16 **56**; 44 **57**; 4, 14 **59**; 14 **60**; 9
Permanence (of form)	**51**; 33 **58**; 20 **59**; 38
Polynomial expansion	**50**; 16 **54**; 10, 40 **58**; 49 **60**; 20
Progression	**50**; 5, 39, 42 **51**; 9, 11 **52**; 12, 30, 42, 50 **53**; 25, 27 **55**; 45 **56**; 25, 36 **57**; 30, 35, 39, 43 **58**; 14, 21, 28, 37 **59**; 12, 33, 37, 39, 49 **60**; 8, 20, 36.
Proportion (see also Variation)	**50**; 1 **51**; 39 **52**; 20, 48 **54**; 11, 28 **56**; 16 **57**; 8 **59**; 4 **60**; 15, 35, 39
Ratio	see Proportion
Roots (of an equation)	**50**; 3 **51**; 16, 26, 47 **52**; 6, 23 **53**; 4, 44 **54**; 24, 25, 41 **55**; 18, 27, 32 **56**; 7, 23 **57**; 2, 27 **58**; 10, 33, 41 **59**; 27, 34, 44 **60**; 1, 11, 28
Sequence	*see* Progression
Series	*see* Progression
Variation	**50**; 11, 19 **51**; 17, 28 **52**; 25 **54**; 3 **55**; 5 **59**; 20

Geometry

Coordinate geometry	**50**; 49 **52**; 5 **53**; 49 **57**; 23, 25 **58**; 7, 22, 27, 50 **60**; 5, 13, 30
Plane geometry	**55**; 48 **60**; 15
Affine properties	

Angle measurement **51**; 12 **52**; 21 **54**; 13, 19, 31, 37 **55**; 2, 29, 33, 44 **56**; 5, 10, 24, 49 **57**; 11, 44, 47, 48 **58**; 15, 29 **59**; 15 **60**; 38

Area **50**; 8, 9, 33, 46 **51**; 2, 3, 10, 25, 48 **52**; 24, 37, 38, 46, 49 **53**; 13, 15, 33 **54**; 5, 8 **55**; 14, 15, 21, 31 **56**; 27, 30, 35, 37 **57**; 7, 25, 26 **58**; 16, 18, 21, 31, 35 **59**; 2, 10, 21, 26 **60**; 4, 6, 7, 15, 21, 37.

Circles **50**; 8, 33, 34, 35 **51**; 46 **52**; 8, 29, 33, 43 **53**; 10, 11, 14, 34, 50 **54**; 9, 26, 39, 43, 46 **55**; 34, 36 **56**; 5, 10, 14, 45, 49 **57**; 46, 47, 48 **58**; 15 **59**; 32, 41, 43 **60**; 5, 6, 7, 12, 35

Classification (of figures) **51**; 27, 29, 36, 46 **52**; 15 **53**; 32 **56**; 26, 29 **57**; 1, 21 **59**; 3, 6 **60**; 12, 13

Congruence **53**; 32 **55**; 48 **57**; 48

Polygons

3 sides **50**; 9, 35, 46, 48, 49 **51**; 27, 32, **52**; 15, 19 **55**; 48 **56**; 24, 26, 38, 43, 49, 50 **57**; 44 **58**; 29, 36 **59**; 36, 40, 43 **60**; 4, 13, 15, 38, 40

4 sides **51**; 2, 3 **52**; 33, 38 **54**; 45 **55**; 48 **56**; 50 **57**; 50 **59**; 22, 40 **60**; 6, 21, 37

n sides ($n > 4$) **50**; 12, 45 **60**; 6, 27

Proportion **50**; 33, 47 **51**; 10, 30, 48 **52**; 19, 24, 27 **53**; 12, 26, 28 **54**; 8, 29 **55**; 14 **56**; 37 **57**; 18, 37, 49 **58**; 42, 47 **59**; 28 **60**; 32

Pythagorean formula **50**; 32, 49 **51**; 49 **52**; 22, 29, 30 **53**; 37, 41, 42, 46, 48 **54**; 35, 47 **55**; 9 **56**; 39 **57**; 46 **58**; 19, 43 **59**; 7, 15, 41 **60**; 40

Ratio *see* Proportion

Similarity *see* Proportion

Solid geometry **50**; 21 **51**; 4, 6, 23 **52**; 41 **54**; 27 **57**; 6, 17 **59**; 1 **60**; 33

Miscellaneous Topics

Logic **51**; 14 **53**; 40 **55**; 11 **56**; 32 **57**; 21 **58**; 44 **59**; 47 **60**; 10

Sets

Algebra of **54**; 2, 18 **59**; 47

Element of **50**; 45 **51**; 31 **52**; 31 **58**; 14, 44, 49 **59**; 19, 48, 50

www.ingramcontent.com/pod-product-compliance
Ingram Content Group UK Ltd.
Pitfield, Milton Keynes, MK11 3LW, UK
UKHW020050200726
13845UKWH00055B/628

9 789324 191762